Cores and Core Logging for Geoscientists

2nd edition

Graham A. Blackbourn

Published by
Whittles Publishing,
Dunbeath,
Caithness KW6 6EY,
Scotland, UK
www.whittlespublishing.com

Distributed in North America by
CRC Press LLC,
Taylor and Francis Group,
6000 Broken Sound Parkway NW, Suite 300,
Boca Raton, FL 33487, USA

Reprinted 2012 on softback ISBN 978-184995-067-1

Original hardback 2009 978-1904445-39-5
USA ISBN 978-1-4398-0116-1

Typeset by Kerrypress Ltd, Luton
Printed and bound in Great Britain by 4edge Ltd, Hockley.

Contents

The following figures can be found in the colour section between pages 72 and 73.

Key to colour section

Preface

Trained geologists logging or interpreting core for the first time seem to fall into (or somewhere between) two camps. There are those who, having worked extensively on outcrop studies (perhaps during a university field-mapping course), rush in under the impression that core should be treated as though it were merely one long thin outcrop of rock. Others are so bewildered at the prospect of dealing with rock in a form quite different from that which they have previously encountered that they hardly know where to start, and may be brought to a halt wondering whether they should begin logging at the top or the bottom of the core. This book has been written to caution the former, and to encourage the latter. Certainly there is nothing formidable about core. Indeed it can be much easier to work with than a rock outcrop. However, it can also present many pitfalls for the unwary, especially when they have little idea as to how a core is cut and recovered, and therefore how it may have been altered from its original state within the rock formation.

No attempt is made to teach geological principles in this book. It is assumed that the reader has some general geological knowledge, together with specialist knowledge of his or her particular field, be it sedimentology, geotechnical engineering, petrophysics or some other discipline. However, methodology is covered in some detail, ranging from basic core handling at the wellsite and beyond, through to logging, interpreting and presenting core data. As such, it is aimed at those who are faced with putting their knowledge of geological principles into practice in a working environment, whether at a senior student or professional level. The book should be of value not only to those who actively work with core, but also to others who use information derived from core studies, who need to be aware of its limitations and scope.

There are always problems when writing a text of this kind in striking a balance between laying down how a thing must be done (even when there are reasonable alternatives), and offering so many of these alternatives that the reader is left with no real guidance as to how the job can best be undertaken. There are many cases where most geologists would agree how a thing should be done (for example, labelling a rock sample in two places, in case one label is lost or damaged), but where few actually do so because they cannot justify the additional time or inconvenience that may be involved. (Laziness, of course, is not a factor!) In many cases there is no ideal method—any approach is inevitably a compromise. I have tried in this book to spell out the advantages and disadvantages of different methods of handling and interpreting core. In many cases I have given specific advice, usually based on personal experience. However, I would have no criticism of the geologist who chose not to heed this advice, so long as this was the result of a conscious decision, rather than mere carelessness.

In the ten years or so leading up to the first edition of this book in 1989, when I was cutting my teeth as a petroleum sedimentologist in the North Sea, the area was changing from a frontier exploration province to one of the major hydrocarbon-producing regions of the world. The high costs of drilling in such inhospitable offshore acreage over this period led to numerous innovations being introduced in both the technology and the methodology of coring, and the handling and analysis of cores. Some of these innovations were experimental and were not pursued, whereas others have become established techniques and are now found around the world. The pace of change has slowed: in the North Sea because it has become a mature province, and around the world perhaps mainly because of relatively low oil prices over the past couple of decades, although at the time of writing they are again testing new highs. This does not mean that there have been no exciting new developments; the huge increase in computing capabilities is almost too obvious an example to cite, but that makes it no less significant. The length of individual cores routinely cut and recovered has steadily increased, so that it is not uncommon to log cores several hundred feet long, and the idea of horizontal coring, virtually unheard of during the 1980s, has now become commonplace.

I might add that, while inevitably written with a certain bias towards petroleum geology, the book was intended to be of interest to a wide circle of geologists working in a variety of industrial and academic disciplines. I have been gratified to discover that it has indeed been taken up by economic, mining and geotechnical specialists, and I am grateful to several of them from around the world both for their kind comments and their words of advice!

Like most geologists, I learnt about core logging 'on the job'. I am therefore grateful to a whole host of friends and colleagues, both from my earlier employment in Britoil and BP, and from working with numerous clients and associates since. They are far too numerous to mention individually, but they know who they are. But I must give my particular thanks to Bob Leppard for making useful suggestions for this second edition, and to Moira Thomson for her many years of patient and mostly uncomplaining hard work and assistance, well beyond the call of duty, and for redrafting most of the figures.

Lastly I once more have to thank my wife, Barbara, still here after all these years and still looking after more and older children, and helping me to keep things in perspective.

Graham A. Blackbourn

1 Introduction

Most geological training in universities is based on surface exposures. This is both necessary and inevitable. It is necessary because geology—the study of the Earth—deals primarily with processes that operate on a large scale. Although geophysical techniques are able to tell us a great deal about the large-scale structure of our planet, rocks can be observed directly over a wide area only by locating good surface exposure. The inevitability of this type of study predominating in our universities arises from the expense involved in obtaining sufficiently large amounts of subsurface sample material covering the wide range of lithologies and structures that would be needed to satisfy the requirements of students.

This emphasis on the study of surface exposure in the academic world is in marked contrast to the workings of industry. Although outcrops are the starting point for most geological work wherever undertaken, few projects would proceed far in the industrial sphere without using information from boreholes. This is the case in a wide range of industries, from geotechnical studies related to site investigation to the search for oil, or for methods of disposing of industrial or nuclear waste. Only the rare professional geologist outside university spends a significant amount of time on outcrop studies, and even where outcrop data are required, these are often available already in the literature.

But although a broad geological training at university is undoubtedly the main requirement for a graduate embarking on a geological career, the change in approach on entering industry can present quite a culture shock. The junior geologist has to learn 'on the job' how to log, say, 250 m of core, with only the experience of several months' field mapping as a guide. Furthermore, there is a whole range of strange jargon—'core-to-log shifts', 'kelly bushings' and 'rabbits'—to cope with. It is important that the geologist is not discouraged, since dealing with cores and logging are quite straightforward, but they need to be undertaken methodically and with care.

Nonetheless, logging and studying cores does improve markedly with practice. It takes experience to recognize all the features in a core caused by the coring process itself, and to distinguish these from original rock properties. In a core, some types of structure stand out much more clearly than they would in outcrop, whereas others are obscured. Indeed, cores of some formations would hardly be recognized when compared alongside their weathered counterparts in surface exposures. The geologist needs to appreciate these differences. Interpretation of cores, where information is effectively packed into a one-dimensional section, is quite different from that of two- or three-dimensional outcrops. Most geological 'models' in the literature are described in at least two-dimensional terms, and the difficulties inherent in interpreting core on the basis of these models can be considerable, and are often underestimated.

The aim of this book is to act as a guide for the trained geologist who needs to apply his or her accumulated knowledge and experience to the description and interpretation of rock sequences recovered by the coring of subsurface boreholes. By describing the limitations and weaknesses, in addition to the advantages, of core in providing geological information, the book should also be of value to those who use data derived from cores, but have not undertaken the logging themselves.

Geologists involved with cores and coring work in a wide range of industries, in which widely differing techniques and terminologies are used. So far as possible, this book seeks to cater for geologists working in any of these spheres, and to concentrate on aspects common to coring, irrespective of the final purpose of the exercise. Inevitably, this can be only partially successful, and many geologists will find techniques described that are different from those they use, and terminology that they do not recognize.

It is in the hydrocarbon industry that coring and core analysis have been most extensively developed, and geologists working in other areas may find that some of the core handling and analytical methods described here appear to go to extraordinary lengths. This is because the cost of retrieving cores from deep hydrocarbon wells can be equally extraordinary (though justifiable), and it is vital to maximize the use that is made of them. Nevertheless, the basic principles of core studies do not change, and it is hoped that the various methods described here from the different sections of industry may lead to a valuable transfer of ideas.

This variation in terminology is exemplified by the simultaneous use of both metric and imperial units. Most measurements in this book are given in SI units. However, drilling in the oil industry is still dominated by American technology, and in some respects—such as borehole and casing diameters, which are invariably quoted in inches—to convert to SI for the sake of uniformity could only cause confusion. In an industry that has been known still to measure oil volumes within the reservoir in 'barrels per acre-foot', there is little prospect of a speedy adoption of uniform standards, and idealism must sometimes give way to pragmatism.

2 Drilling and coring methods

Wells have been sunk for water, brine and even oil for thousands of years. Until the early nineteenth century they were dug by hand, but the development at that time of percussion ('cable-tool' or 'churn') drilling enabled the operator to remain on the surface. In some developing countries, however, hand-digging remains the most common method of sinking water wells.

Except in shallow, unconsolidated formations, the percussion drill is capable of recovering only disordered chippings of the rock, which are suitable only for general lithological description. It is said that the first coring device, capable of recovering intact samples, was designed by a driller in Holland in 1908, following the advent of rotary drilling. This device consisted of a steel tube positioned in the middle of the drill bit. Development of the first effective tool, however, is credited to J.E. Eliot of the United States, who, between 1921 and 1925, introduced a core barrel with removable core head, core catcher and stationary inner barrel. Despite numerous refinements, the basic coring tool has not changed in principle to the present day.

Coring is just one of many operations undertaken by the modern driller. Since any understanding of the advantages and limitations of coring requires knowledge of general drilling techniques, a short summary is provided here. It must be noted, however, that drilling is a complex multidisciplinary craft, and this section can deal with only a few of the more common methods.

2.1 Percussion drilling

Percussion drilling works by repeatedly dropping a heavy metal bit, usually a solid-steel chisel, down the hole (Fig. 2.1). This chips fragments from the bottom of the hole, which are periodically removed in a 'bailer' and may be inspected at the surface. The 'bailer', which comprises a heavy steel tube with a non-return valve at the base, also empties the hole of any water that may seep in from the sides. In soft, unstable formations a steel tube or 'shell', with a cutting shoe on the bottom, is used (Fig. 2.1). This serves to drill into the formation, in addition to acting as a bailer. Percussion drilling has been used to depths of 3000 m, but about 1000 m is a more average maximum capability, and in practice in most parts of the world it is now used primarily for site investigation work at depths of up to around 60 m. Only vertical wells can be sunk by this method, with a

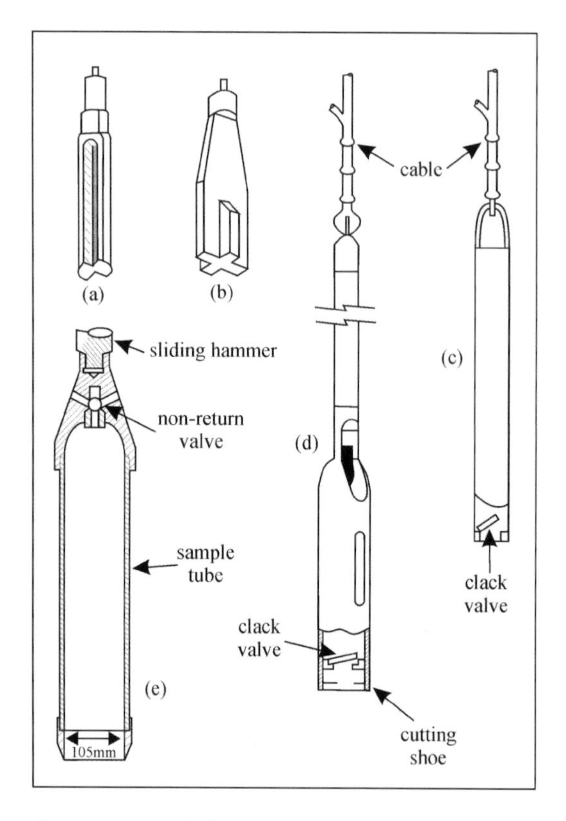

Fig. 2.1 *Tools used in percussion drilling: (a) heavy steel drilling bit, and (b) chisel;(c) bailer; (d) shell for penetrating soft lithologies; (e) standard open-drive (U100) sampler.*

diameter usually between about 150 and 600 cm. Percussion drilling will not be considered at length in this book, since in general it cannot be used to take core. The exception is in shallow, unconsolidated formations and drift deposits, for which a range of sampling tools have been devised. These are mostly variations on a hollow open-ended cylinder, the most common of which is the 100 mm (nominal) internal diameter, or U100, open-drive sampler (Fig. 2.1). The U100 sampler can be machined to accept a metal or plastic liner, to facilitate removal of the core intact. A liner that is split along its length is an additional help. The U100 assembly is lowered to the bottom of the borehole, and penetrates the formation by repeated blows from a slide hammer or drop weight. A count of the number of blows required to penetrate 450 mm (i.e. the full length of the sampler) provides an indication of the hardness of the formation. On removal from the borehole, the liner or U100 tube is sealed with endcaps and may be stored in plastic tubing or bags until examination of the sample can take place.

Some types of sampler contain a spring mechanism or 'core catcher' to help prevent the sample from falling out of the bottom of the tube. However, it is the natural cohesiveness of the formation that is largely responsible for successful recovery. A cohesive sample (for example wet clay) will not immediately disaggregate, and will also tend to stick to the inside walls of the sample tube. Because of this, recovery of non-cohesive material (for example loose sand or gravel) by percussion samplers is difficult, although various modifications have been developed for this purpose, with varying success.

Even where the recovery is good, the essentially violent sampling technique usually results in significant sample damage. This damage is commonly concentrated, however, in the upper part of the sampler, where disturbance caused by the slide hammer is most intense. It is possible to join two samplers 450 mm long by a collar and, having penetrated 900 mm into the formation, to discard the upper section to leave a single, relatively intact, core.

Even in the shallowest wells the borehole walls are liable to cave in, and the resultant 'cavings' fall down the well. This can lead, at best, to contamination of samples retrieved from the bottom of the hole and, at worst, to total collapse of the formation into the well. This would usually require abandonment of the well, with loss downhole of the bit and cable. These problems are usually avoided by progressively lining or 'casing' the well, as it is drilled, with steel (or in some shallower wells, plastic) tubing. The initial section of casing, which rises above the surface and helps to guide the various drilling apparatus into the hole, is known as the conductor pipe (Fig. 2.2).

Because the cable and bit of a percussion drill need to fall freely under gravity, it is not normally possible to 'weight' the hole with dense drilling muds and thus prevent the flow of formation fluids into the well. Moderate amounts of fluid are removed with the bailer, but there is always a danger that drilling into a high-pressure formation may result in the fluids rushing up the well-bore and emerging violently at the surface as a blowout. If the fluid is water, the results would be dangerous and inconvenient. With gas or oil, one would be fortunate to avoid a major catastrophe. Despite having been the norm in the latter half of the nineteenth century, percussion drilling is no longer used in the hydrocarbon industry.

The advantage of the percussion over the rotary drilling method is its relative simplicity and thus cheapness. For this reason, it tends to be used in relatively deep vertical wells where core is not required (for example some water wells, particularly in developing countries) and in studies of soft near-surface lithologies and overburden (such as site investigation). Marine survey and research ships also use varieties of percussion sampler, such as piston corers and vibrocorers, to obtain soft seabed cores.

2.2 Rotary drilling

A rotary drill essentially comprises a bit that is rotated at the end of a steel tube. The rotational torque is applied to the tube, known as the 'drill string', at the surface. The simplest form of rotary drill is the hand auger. This is little more

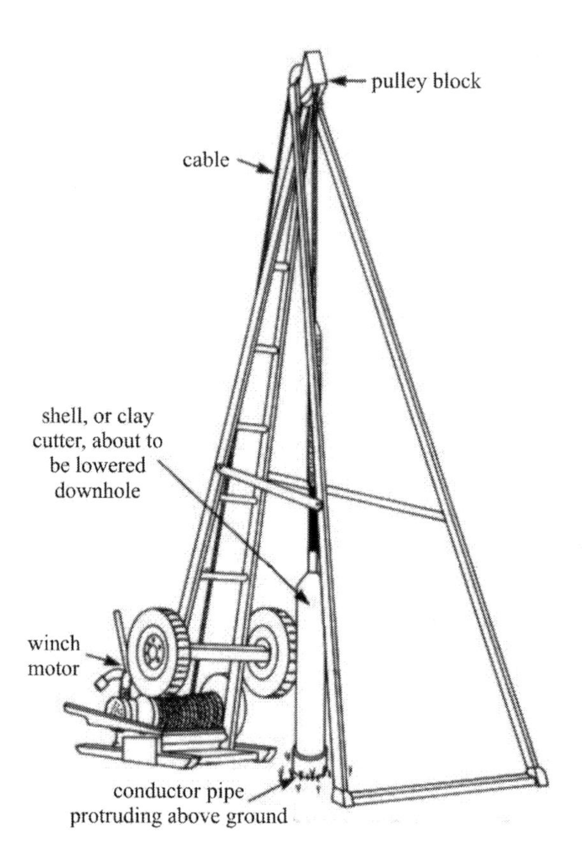

Fig. 2.2 *Small percussion rig of a type widely used for site investigation.*

than a large corkscrew that is twisted into the soil to a certain depth and then pulled out without twisting.

In theory, an ordered vertical profile of the soil should be recovered within the flights or 'thread' of the auger. In fact the hand auger, despite its obvious limitation as to depth of penetration, is a very effective tool for sampling near-surface cohesive soils and clays, and any small fragments of shallow bedrock that may be included within them. There are hand augers available that, under suitable soil conditions (self-supporting yet soft material), can drill holes up to 200 mm in diameter and, using extension drill rods, up to about 5 m deep. A small auger has the advantage of being sufficiently light and compact to be carried into the field in a rucksack.

Mechanical auger rigs are widely used for rapid sampling of soft and cohesive formations at shallow depths, and for drilling in the initial stages of a deeper well (Fig. 2.3). The most common design consists of a spiral blade welded onto a central shaft, which can be extended by fitting additional sections typically 1 m long. They can be operated from freestanding rigs, or

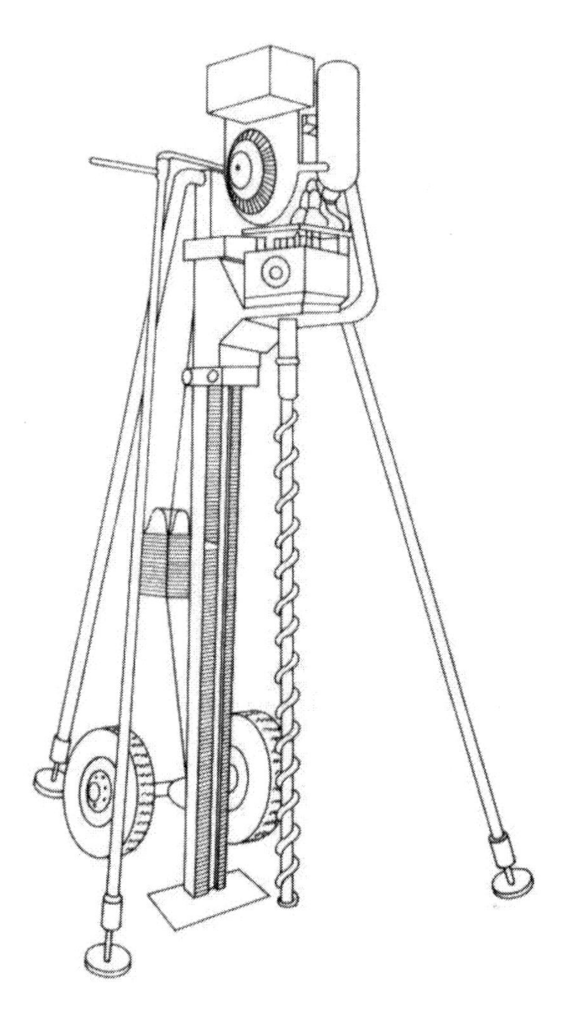

Fig. 2.3 *Small continuous-flight power auger rig.*

mounted on a truck or crane. The largest varieties can bore holes over 1 m across.

Relatively undisturbed core samples can be recovered using a hollow-stem auger, in which the central shaft is a hollow tube with a diameter typically of 75–125 mm. A core of formation passes into the shaft as penetration of the auger proceeds, and may then be pulled up and removed intact for examination. Alternatively, the hollow stem may be plugged at the bottom while the auger itself is in operation, and then a sampler, or even a core-cutting assembly, lowered down the stem to take relatively undisturbed small samples.

The first patents issued for rotary drilling comparable to modern methods, in which a drill bit is rotated on the bottom of the borehole, were held by Robert Beart in England in 1844, but the technique took some time to become established. By the 1890s it was becoming more widely used, and the value of drilling mud not only to lubricate and cool the bit, but also to remove rock fragments and to seal the borehole wall, had been recognized (Section 2.4). A major impetus for rotary drilling, especially in oil exploration, came in 1901 when the Hamill brothers successfully used a rotary drilling rig to drill the second well on the prolific Spindletop field, Texas. The first well had been drilled with conventional percussion tools and had to be abandoned owing to formation caving. Nonetheless, it was still percussion drilling four years later in 1905 that discovered the Glenn Pool field, at a depth of nearly 1500 ft, leading to the Oklahoma oil boom.

The introduction in 1908 of the rolling cutter rock bit, which drilled by means of two rotating cutters, was a further advance. Numerous types of bit are available nowadays for rotary drilling. When not coring, the most common type is the tri-cone roller bit (Fig. 2.4), or variations of it, first used in 1933. This has three rotating cones set on high-grade bearings, which rotate independently of each other. The cones are set with teeth, which may be made of the same material as the cones, or may be of a harder material (such as tungsten carbide) for use in harder formations. The three cones are mounted so that the teeth of each clean the channels of the other two. The whole tri-cone bit is rotated at the end of the drill string, and the teeth gouge away or chip at the formation. The resultant chippings or 'cuttings' are carried to the surface in the flow of drilling mud, where they may be inspected.

For harder formations, diamond bits may be used. Commercial diamonds are set in the matrix of the bit, which is rotated and the formation penetrated by abrasive action. Other bits, such as polycrystalline diamond bits, are designed for

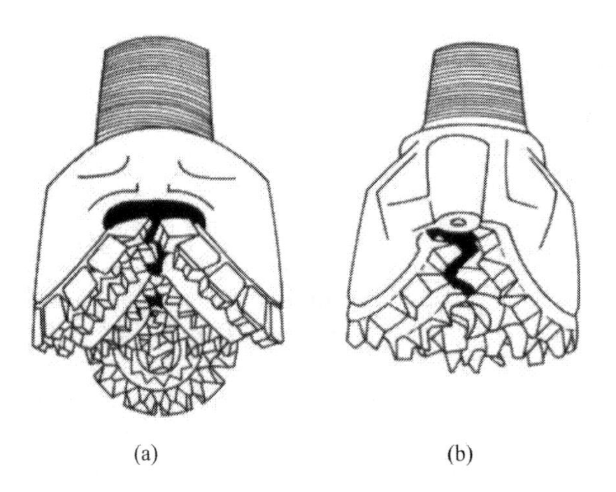

Fig. 2.4 *Tri-cone roller bits, designed for (a) hard, and (b) soft formations.*

relatively soft, cohesive formations and scrape at the rock. Unlike the tri-cone bits, these bits have no moving parts.

During drilling, water, mud or some alternative fluid (Section 2.4) is pumped down the hollow drill string, to emerge under pressure through nozzles in the bit. It then flows up the well, between the drill string and the borehole wall or, if in place, the casing. At the surface the mud is allowed to settle in a tank or pit, having sometimes passed first through a vibrating mesh screen known as a 'shale shaker', from which the geologist can retrieve cuttings for inspection. In the absence of a shale shaker, a sample of mud can be collected from the mud channel in a container and cuttings allowed to settle out for sampling. Controlling the composition and flow of the mud is one of the most critical operations in drilling a deep hole. The mud not only removes rock cuttings and cavings from the well and keeps the bit cool; it is also kept at a sufficient density to prevent any high-pressure formation fluids from escaping uncontrollably into the well and causing a blowout.

In some small-diameter boreholes, up to several hundred metres in depth, compressed air is used as a circulating medium rather than mud or water. This reduces contamination of the borehole wall and of samples.

With the exception of some small rotary drilling rigs designed for penetration of several metres at most, the drill string consists of a number of columns of drill pipe, which are added on successively as drilling progresses, much as an old-fashioned chimney sweep adds sections to his broom handle while pushing it up the chimney. Also like some sweeps' brooms, the sections screw into one another using taper thread joints. The topmost section of drill pipe is attached to a square (or sometimes hexagonal) section steel member called the 'kelly' (Fig. 2.5). Rotational motion is transferred to the kelly by the rotary table, a horizontal platform with a square or hexagonal hole in the centre (the kelly bushing) through which the kelly passes. The rig motors turn the rotary table and kelly, and the rotational movement is transferred down the drill string to the bit. As the bit moves deeper, the kelly drops through the kelly bushing under the weight of the drill string. Because the kelly slides freely through the kelly bushing, the weight on the bit at the bottom of the hole remains constant. Once the length of the kelly has slid through the kelly bushing, drilling stops so that another length of drill pipe may be attached to the drill string.

In addition to its functional importance, the kelly bushing is a convenient datum from which to measure downhole depths. The use of this datum may be indicated by the letters 'KB' or 'BKB' (below kelly bushing), as in '3000 m BKB'. Note that the rotary table, which is also sometimes quoted as the datum ('BRT'), is at the same level as the kelly bushing (Fig. 2.6).

As an alternative to the kelly and kelly bushing, a method sometimes used to rotate the drill string is the use of top drive. The top section of drill pipe is connected directly to a rotary head, which is driven by a motor mounted immediately adjacent to it. During drilling, the rotary head and motor are

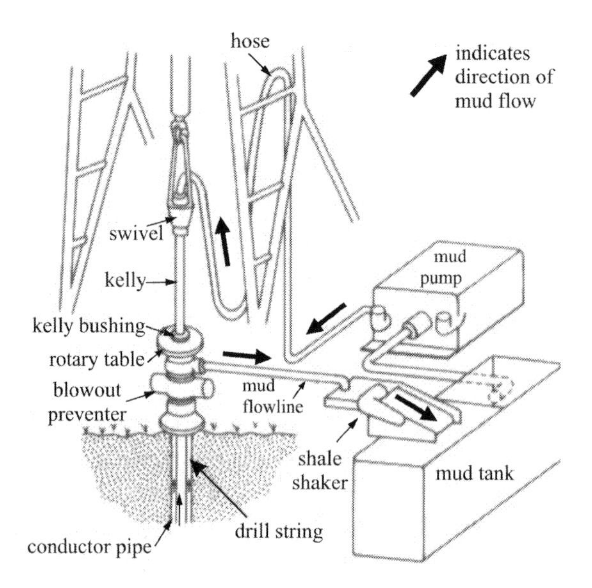

Fig. 2.5 *Main components of a typical rotary drilling rig on a land-based hydrocarbon exploration well.*

allowed to slide down from the top of the rig to the drilling floor attached to the drill string. The rotary head is then detached from the drill string and hauled back to the top of the rig, while another length of drill pipe is added so that drilling can recommence.

As with percussion drilling, rotary-drilled wells are normally lined with casing if they are to exceed the shallowest of depths. Steel casing is widely used, but plastic or fibreglass are suitable for shallower wells. In the oil industry, the top-hole section of a deep well is commonly drilled with a 36-inch-diameter bit and lined with a 30-inch-diameter casing. As greater depths are reached both bit and casing diameters are reduced, and the bottom hole may be drilled with an 8.5-inch bit and lined with 7-inch casing.

2.3 Non-conventional drilling methods

A brief mention should be made of several less conventional drilling methods that have been introduced, especially in the hydrocarbon industry. The first of these is the downhole motor, or turbo drill. The downhole motor is powered by the circulating drilling mud. It consists essentially of a turbine, which drives the drilling bit. With the mud circulating, drilling progresses without the need for the drill string to rotate. Drilling with the downhole motor is often significantly faster than with a conventional tri-cone bit. The major disadvantage, especially for geologists, is that the cuttings produced by the bit are small and have often

Fig. 2.6 *The (square-section) kelly bushing within the (circular) rotary table in the centre of the drill floor. The rotary table is here stationary, prior to a coring run. (Photo courtesy of Diamant Boart Stratabit.)*

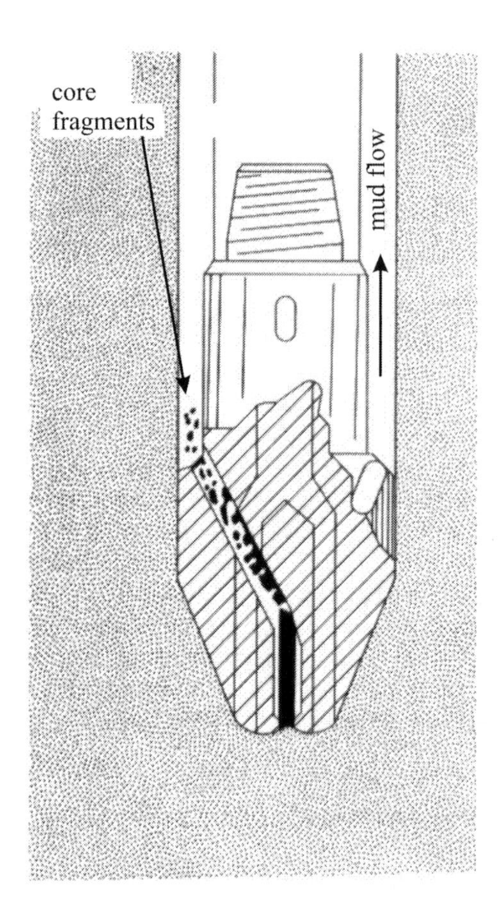

Fig. 2.7 *Schematic diagram of a core ejector bit penetrating the formation.*

suffered significant thermal alteration. The cost-effectiveness of the downhole motor, resulting from its rapid penetration, needs to be weighed against the considerably poorer geological information that is likely to be obtained. For this reason, turbo drills are usually appropriate only when the chief aim is to 'make hole', rather than to obtain sample material from a well.

A core ejector bit is a form of diamond bit that cuts a small-diameter core at its centre (Fig. 2.7). There is no core barrel, and the core is designed to break up and be carried to the surface with the conventional cuttings. The resultant samples are, in effect, large cuttings rather than cores, but they do have the potential to provide more lithological information than conventional cuttings.

2.4 Drilling muds

It has already been stated that control of the composition and flow of the drilling mud is one of the most critical operations in drilling a deep hole. It is not intended here to discuss the details of mud technology, but to outline a few factors that will affect the rock samples and sampling programme.

The earliest muds were a simple mixture of water and clay (or 'gumbo'), which, during drilling, inevitably became mixed with rock flour and cuttings. The mud is mixed at the well site, according to conditions downhole. Historically this has been carried out by driving cattle through a pit containing clay and water, but the process is now mechanized. Bentonite is the most widely used clay type. During the 1920s it was found that adding powder of dense minerals, such as barite, to the mud would raise the mud density and thus prevent the movement of high-pressure fluids from the formation into the well. Barite is still an important constituent of many drilling muds. Modern 'muds'—the name is now used for a wide variety of substances—may include a range of natural and artificial components, and in addition to simple suspensions of solid matter in liquid they include colloids, emulsions and foams.

Because the pressure of mud at any depth downhole is designed to be greater than that of formation fluid at the same depth, there is always the possibility that it will flow into the formation along permeable zones or fractures rather than return up the well-bore.

This will inevitably happen to some extent, but fortunately the formation usually acts as a filter. The fluid portion of the mud ('mud filtrate') enters the rock, while the solids are filtered out and accumulate on the borehole wall as a 'mud cake'. This mud cake is impermeable, and usually prevents further penetration of mud filtrate into the rock. Because mud filtrate will not enter an impermeable formation, no mud cake will form. Thus the thickness of mud cake and depth of penetration of filtrate are an indication of the porosity and permeability of the rock. These factors can be interpreted from electric wireline logs (such as the resistivity log, Section 2.8). This partial penetration of mud into the formation is thus beneficial to the interpreting geologist. On the other hand, by severely reducing the porosity and permeability of the formation in the vicinity of the well, it can cause major problems for the production of hydrocarbons or water from the well.

Sometimes, because of excessive 'lost circulation' of mud due, for example, to large open fractures in the rock, special measures need to be taken. This usually entails the addition of fibrous, flaky or granular materials to the mud to 'stop up' the fractures. The list of additives used is certainly imaginative, and includes peat moss, cellophane, feathers, corn cobs, walnut shells and ground tyres. It is important for the geologist to be aware of the range of items added to the mud that may turn up in the geological samples.

Numerous other mud additives are used to solve particular problems. These include salts (drilling through evaporites with a water-base mud would pose obvious difficulties unless the mud was already saturated in the relevant salts), various ground-up minerals and clays, and organic fluids. Lignosulphonates are used to prevent the swelling of water-reactive clays. They have been known to contain recognizable spores and pollen derived from natural lignite, which could turn up in geological samples and give spurious biostratigraphic ages. They could also be responsible for invalid results of analyses for carbon content, as could graphite, which is sometimes used as a lubricant.

One of the most controversial mud types is oil-base mud, which uses oil rather than water as the main fluid. There is no doubt that the performance of oil-base muds is superior to that of water-base muds in certain conditions, especially in the drilling of deep, hot wells, and their use is therefore often recommended by the driller. Oil-base muds can greatly improve penetration rates of downhole motors. They also maintain the hole in a significantly better condition, prevent the swelling of water-sensitive clays, and reduce corrosion in the well. Unfortunately, they inevitably hinder the interpretation of hydrocarbon shows in a well, particularly when thermal cracking of the oil by the bit produces a tarry residue, which may contaminate the cuttings. Many electric logs are ineffective, or their interpretation is hampered, where oil-base mud has been used. Conventional resistivity logs, in particular, cannot be used, and have to be replaced by induction logs (Section 2.8). In addition, oil-base muds can be unpleasant to handle, and are mildly toxic.

In general, oil-base muds are effective and economic for drilling oilfield development wells, where the emphasis is on producing a good-quality hole rather than on interpretation. They are also justifiably used in deep exploration wells, where there is otherwise a significant chance of failure of the well. In most other situations the importance of gaining good geological data outweighs the advantage of using an oil-base mud.

It is, of course, important for the geologist to be aware of the problems and ambiguities that may arise when an oil-base mud has been used to drill a well.

2.5 Coring

When a formation is to be cored, the normal bit is replaced by a coring bit and core barrel (Fig. 2.8). The coring bit is an annular cutting ring, usually studded with diamond or tungsten carbide inserts, which fits onto the bottom of the core barrel (Fig. 2.9). The core barrel, in fact, usually consists of an inner and an outer sleeve. The outer sleeve transfers weight and rotation to the bit, while the inner sleeve is mounted on roller bearings and may remain stationary. The drilling fluid passes down between the inner and outer barrels. The core, as it is cut, passes up into the inner barrel, into which it should fit closely, but not tightly. Because this barrel remains stationary, the core is relatively undisturbed. Coring bits are smaller and much more delicate than tri-cone and other rotary bits, so coring must proceed slowly and with only a minimum of pressure on the bottom.

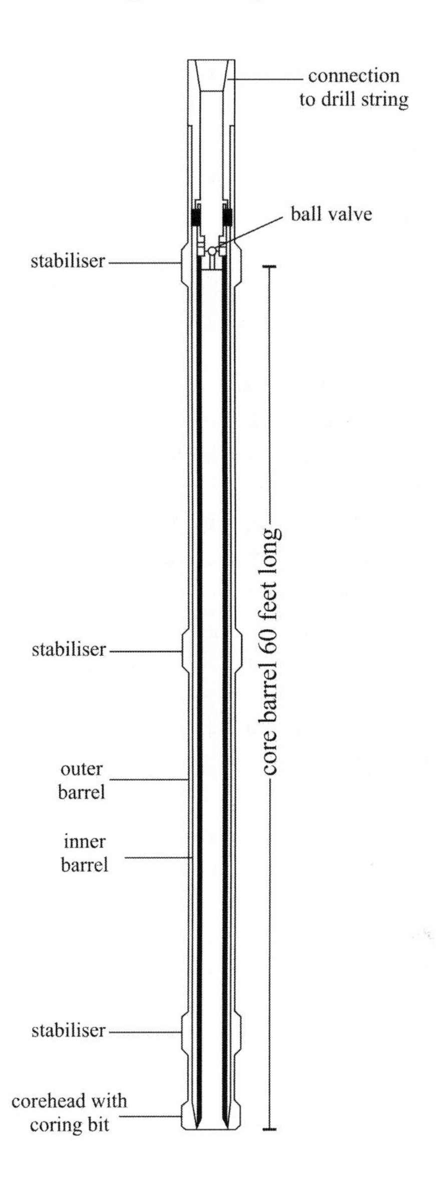

Fig. 2.8 *Diagram of a standard core barrel of the type used in the hydrocarbon industry.*

On completion of coring, the core barrel is hauled up. The core itself is prevented from falling out by a steel spring device known as the 'core catcher' (or core lifter). Thus the core is broken from the formation and raised to the surface.

As an alternative to the need for removal of the entire drill string to recover the core barrel, the technique of wireline core drilling is sometimes used. The internal diameter of the drill string is sufficiently great to allow the core barrel to pass through. Once the core barrel is full, it is retrieved with a special tool with a

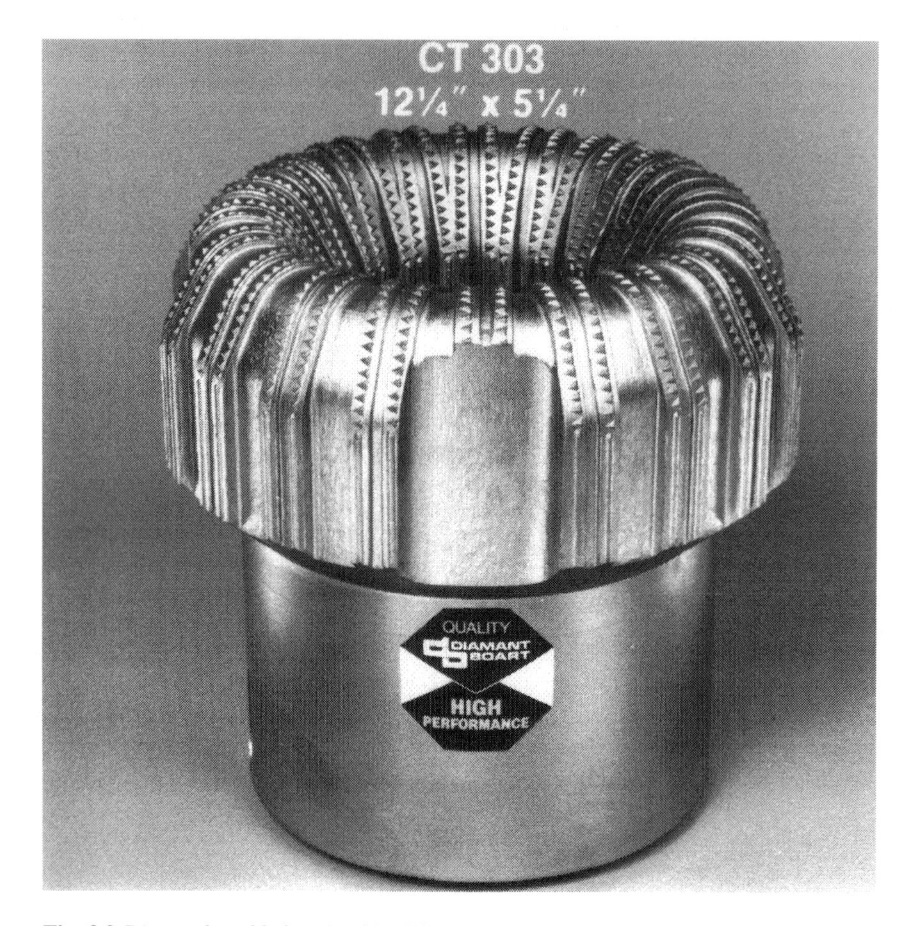

Fig. 2.9 *Diamond-studded coring bit. (Photo courtesy of Diamant Boart Stratabit.)*

bayonet fixing, which is lowered down the inside of the drill string on a cable. The bayonet fixing engages with its counterpart on the top of the core barrel, allowing the barrel to be withdrawn to the surface through the drill string.

Rock cores typically range in diameter from about 20 mm to 150 mm, and sometimes more. Core barrels, and thus individual cores, generally range in length from about 1.5 m to 30 m, although modern drilling techniques using top-drive drilling (Section 2.2) have enabled individual cores substantially above 100 m in length, indeed as much as 300–400 m, to be recovered by connecting a number of shorter core barrels together in series.

The conventional coring method described above is used while drilling the well. If a particular horizon needs to be cored, such as a mineralized zone, or the top of a reservoir sand, its proximity must be predicted in advance. The core barrel is then run downhole to cut core over the required interval. In shallow wells it is possible to core continuously and thus guarantee recovery of zones of interest, but in deep wells this is usually both prohibitively slow and expensive.

Picking the correct coring point, where a particular stratigraphic interval is required, is not always easy. In deep wells, geophysically based estimates of the depth to a certain horizon are not usually sufficiently accurate. In a known and variable succession it is sometimes possible to recognize from cuttings a distinctive lithology that occurs close above the required coring point. Similarly, MWD (measurement while drilling; Section 2.8) used in a known succession can indicate the proximity of the required coring point. Alternatively, there may be some palynological or micropalaeontological marker horizon above the coring point, which can be spotted by a wellsite biostratigrapher.

An obvious technique that is quite widely used in petroleum wells in the USA is to drill ahead until the coring target is penetrated (either completely, or at least its upper surface), and then to raise the drill string and sidetrack the well from a little above the target horizon. The new well-bore is made to run close to and parallel with the previous hole, so that the selected coring point can be calculated with precision. The core barrel can then be run into the well exactly when required, although at the expense of additional rig time.

The use of such specialist or costly techniques can normally be justified, however, only when the recovery of a particular horizon is critical. It is an observable fact that attempts to core through, say, a reservoir sandstone often omit the first few feet, because the core barrel was run in the hole only on recovery of the first sandstone cuttings from the drilling mud. This is unfortunate, but is only one of many examples where some measure of data quality has to be sacrificed in the face of economic constraints.

2.6 Special coring techniques

2.6.1 Scribed and orientated cores

By setting a ring containing three inward-pointing V-shaped projections (known as Hugel scoring knives) into the coring bit shoe, the core entering the core barrel is marked with three continuous grooves, or scribe lines. Two of the knives are set at (say) 90° apart and the third is (usually) centred on the opposite side and is thus 135° from each of the others. This third knife cuts the 'primary' or 'reference' groove. The scribe lines may subsequently be used to help reconstruct the core pieces to their original relative orientation.

Scribe lines are sometimes cut simply to assist such reconstruction of the core. However, they also perform an important function in combination with the core orientation tool. Various methods exist for enabling the original orientation of cores to be recorded. This ability is valuable wherever the alignment of some feature in the formation needs to be discovered, such as in studies of fracture patterns, sedimentary dip or directional permeability. The most widely used system in deep wells consists of the magnetic multishot instrument, which sits on the top of the core barrel and bears a known angular relationship to the scribing knives. In simple terms, the multishot instrument consists of a magnetic compass and a camera. The camera photographs the compass every few minutes during drilling, and thus keeps a record of the orientation of the core barrel. Since the scribing lines have marked the position of the core barrel relative to the core itself, it is possible to use the multishot survey data to calculate the original orientation of the core in relation to magnetic north.

Early orientation tools were unpopular with drillers because the coring and mud circulation had to stop before each photograph was taken. Apart from the time factor, this frequent stopping and starting significantly increased the likelihood of drilling difficulties arising. More recent tools, however, are able to operate while drilling proceeds. As a result the cutting of orientated cores is now a much safer, cheaper, and thus more common operation.

2.6.2 Sleeved cores

When cutting through a soft or friable formation, the core is often held safely in the inner core barrel, only to be disaggregated or damaged during removal from the barrel at the well site (Section 3.2.2), or in transport to the laboratory. Such damage can be considerably reduced by the use of a triple-tube core barrel. The innermost tube (the ‘sleeve’ or ‘liner’) often used to be made of rubber, although plastic, fibreglass or aluminium are now more common. The sleeve sits within the inner core barrel in such a way that the core enters it on being cut. Sometimes the sleeve is split along its length to facilitate subsequent removal of the core. At the wellsite the core, still protected in its sleeve, is removed from the core barrel (Fig. 2.10) and, having been cut if necessary into manageable lengths, is sent to the laboratory (Fig. 2.11).

2.6.3 Pressure and sponge cores

Pressure and sponge cores are cut in such a way as to preserve, as far as possible, the original formation fluids in the rock. They are usually taken for reservoir engineers and petrophysicists in the oil industry, rather than for geologists. By the time the geologist has access to the samples, they may be treated like any other core.

2.6.4 Horizontal cores

Hydrocarbon production wells and occasional exploration wells are being drilled increasingly frequently at angles that are highly deviated from the vertical. In the extreme case the well at depth is drilled horizontally, parallel to the surface. The most frequent reason for this is to maximize production by following a particular thin, hydrocarbon-bearing horizon, sometimes for a considerable distance. Occasionally, along dipping beds, the well may even follow a slightly up-dip trajectory. Drillers were initially reluctant to cut cores from horizontal well

Fig. 2.10 *Fibreglass core barrel laid out on the rig floor. (Photo courtesy of Diamant Boart Stratabit.)*

sections owing to the considerable technical difficulties, but 'horizontal cores' are now becoming more common. To the wellsite geologist, they have to be handled and marked up just like any other core. During logging, however, a mental adjustment needs to be made to allow for a core that explores the lateral variation within a bed, rather than the vertical succession of a sequence of beds,

Fig. 2.11 *Core, still in its fibreglass protective sleeve, but cut into 1-m lengths and with the ends sealed, stacked on the drilling platform and awaiting transport ashore. (Photo courtesy of Diamant Boart Stratabit.)*

and 'depth' measurements obviously have a different significance (Section 4.3.2).

2.7 Sidewall cores

The requirement to pick coring points on a rather 'hit or miss' basis is one practical problem with the conventional core. Another is that, although coring can give good-quality data over the cored interval, even a major coring programme in a deep well will recover only, say, 300 m out of 3000 m, the constraint being the excessive cost of such coring.

Sometimes, good-quality data are required from points over the entire depth of a well. Biostratigraphic analyses, for example, although commonly carried out using cuttings, can be considerably enhanced by the addition of good-quality samples from precisely known depths.

Various devices have been developed to take samples from the borehole wall after the well has been drilled. The tool in most regular use is the percussion sidewall core gun, also called a 'chronological sample taker' or CST. This is basically a heavy steel bar that is run down the hole on a cable (Fig. 2.12). The bar holds a number of small core barrels, about 1 inch in diameter and attached to the bar by a length of steel wire. A small explosive charge, triggered from the surface, shoots the barrel out of the gun and into the borehole wall. When the gun is moved up the hole, the barrel is pulled out of the formation containing, with luck, a 1-inch diameter core, which may be up to about 1.5 inches long.

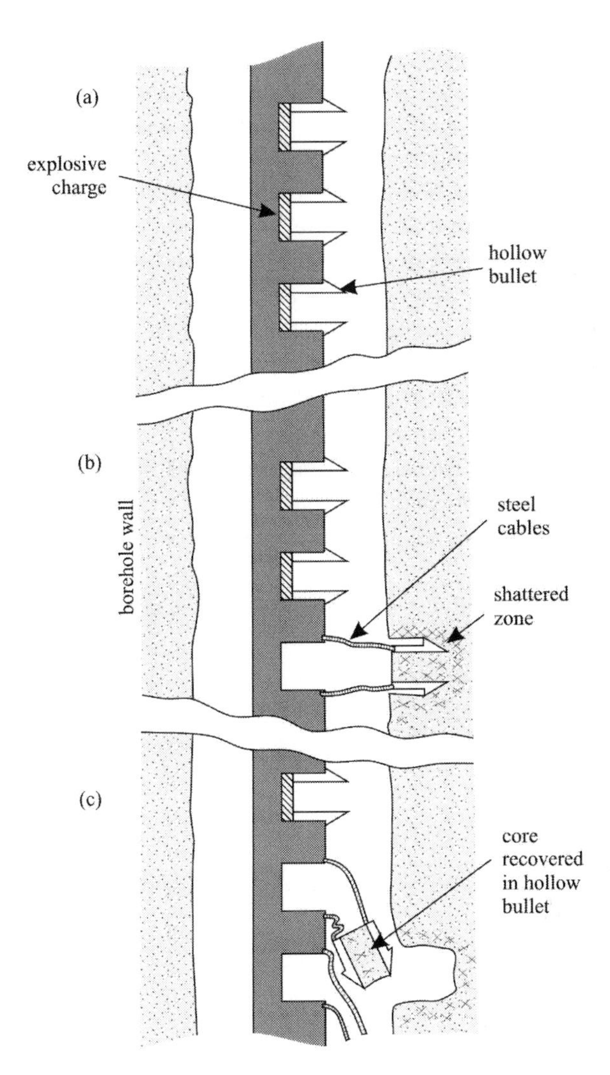

Fig. 2.12 *(a) Sidewall cores are cut from the formation by hollow 'bullets', initially set in the steel body of the tool, each with its own explosive charge. (c) When the tool reaches a required sample depth, a command from the surface detonates one of the charges, and the bullet is shot into the formation. (c) As the tool is lifted up the well, the bullet, attached by steel wires, is pulled from the formation and recovered at the surface.*

The sidewall core gun is usually run after wireline logging of the borehole. The logs may then be used to pick sampling points. Samples for biostratigraphic analyses, for example, may be taken on a gamma-ray peak, which could indicate a shaly lithology (Section 2.8).

Percussion sidewall cores are very useful for sampling uncored lithologies and for biostratigraphic analyses. Because they can be taken after wireline logging, they may be used to solve problems posed by the logs, such as the positioning of an oil–water contact. It is important to note, however, that the forceful method of recovery usually results in a pervasive shattering and alteration of the rock texture.

Sidewall cores are therefore not usually suitable for detailed petrographic analyses, and determinations of rock strength, porosity, permeability and the like are liable to be highly unrepresentative of the original formation.

Two other wireline sampling tools should be mentioned. One has a small diamond coring bit, which extends from the body of the tool and can cut a 1-inch plug, similar in size to a conventional percussion sidewall core. It is most effective in relatively well-indurated formations, and its use is rather more expensive than that of the sidewall core gun. However, the samples recovered do not suffer the same mechanical shattering, and appear to give valid permeability and porosity results on analysis. This technique is therefore useful when better-quality samples are required than those normally recovered by percussion sidewall-core sampling.

Another wireline sampling tool consists of two diamond cutting wheels that penetrate the sidewall and move down the formation to produce a long triangular-sectioned slice of rock, which is retrieved in a sampling chamber in the body of the tool. These continuous triangular sidewall cores may be 3–5 feet long and 1.5 inches wide. This particular tool, however, is not widely used.

2.8 Electric wireline logging

This section is included here since, although not **a** drilling method, wireline logging commonly constitutes an integral part of the drilling programme, especially in deeper wells.

When the first electrical logging device was used in 1927 by Schlumberger and Doll, the technique was known as 'carottage électrique', or electrical coring. The development and interpretation of wireline tools have advanced considerably since then, and they can provide very detailed information on the nature of the rock succession penetrated by a well. Nonetheless, geologists (and most log analysts) would agree that electric logs are not, and will not be in the foreseeable future, a complete alternative to conventional cores. Reliable data on, for example, rock permeability and strength cannot be constructed from logs alone.

Wireline logging tools are run in the hole after a well, or section of a well, has been drilled, but usually before casing has been set. (There are some types of tool designed for use through the casing.) In addition, measurement-while-drilling (MWD) systems attach to the bottom of the drill string and provide more limited data from just above the bit while drilling is in progress. They send data to the surface by relaying it in the form of acoustic pulses through the mud system.

A short summary cannot do justice to the range of tools available, or to the interpretation of the logs produced. The reader is therefore referred to the specialist books on the subject (see bibliography). A brief mention will be made here, though, of some of the most common types of log, and of their major applications.

The 'spontaneous potential' (SP) log has been widely used in the past, and is still common in some parts of the world, such as the former Soviet Union. It records the potential difference between an electrode on the downhole logging device and one connected to earth at the surface. It can be used only with electrically conductive mud. The SP response can depend on numerous factors, but in most permeable formations it reflects primarily the salinity difference between the formation water and drilling mud.

Over the years, the SP log has been used to measure a wide range of parameters, but it rarely gives unequivocal results. Despite being vigorously defended by some stalwarts, for most geological purposes it has been superseded by other logs.

Resistivity logs measure the electrical resistivity of the formation. The most primitive types simply measure the current flowing between two electrodes on a sonde raised up the borehole, although numerous refinements are used to focus the current, which enable resistivity to be measured at various distances into the formation from the borehole wall. Where the drilling fluid is non-conductive (for example freshwater or oil-base muds), an induction log is used, which measures an alternating current induced in the formation.

Most types of solid rock are highly resistive, although shales are usually of low resistivity. However, the major factor affecting resistivity measurements is the formation fluid. Fresh water, oil and gas are highly resistive, whereas brine conducts electricity to a degree depending on its salinity. Because of this, the resistivity log can differentiate hydrocarbon-bearing zones from those containing saline formation water. Furthermore, drilling fluids penetrating the formation affect the resistivity of the formation. Since the degree of mud filtrate invasion is controlled largely by the permeability of the formation (Section 2.4), the resistivity log is one of the better tools for indicating permeable zones.

The gamma-ray log is a measure of the natural gamma radiation emitted by a formation, which in sedimentary successions tends (but not infallibly) to relate to their clay content. 'Clean' formations, such as limestones and quartz-rich sandstones, have a low radioactivity, unless the formations are rich in dissolved radioactive salts. Radioactive elements, however, especially the radioactive potassium isotope ^{40}K, tend to accumulate in clays. Most natural gamma radiation in the Earth, apart from potassium, is emitted by isotopes of uranium and thorium. In addition to clays, other potassium-rich minerals, especially some feldspars, micas and evaporites, give a high gamma-ray response.

The basic gamma-ray log measures total intensity, whereas a more sophisticated version, the gamma-ray spectrometry log, measures additionally the energy levels of the gamma rays, from which the concentrations of radioactive potassium, thorium and uranium can be determined.

The neutron tool bombards the formation with neutrons from a radioactive source. Hydrogen ions emit gamma rays on being struck by neutrons, and this radiation, which is proportional in intensity to the amount of hydrogen present, is measured by the logging tool. As hydrogen is usually present in formation fluids (whether as water or hydrocarbons), but is not abundant in minerals, the neutron log gives an indication of porosity. There are of course various corrections that have to be applied, depending on such factors as hole size and lithology, before an actual figure for porosity can be generated from the log data.

Variations in hole diameter have a significant effect on log responses, and also give an indirect indication of a number of lithological features such as hardness and the degree of fracturing. A caliper device is usually run as part of a logging programme in order to measure and record these variations.

The density log measures the gamma rays produced from a formation that is itself bombarded by gamma-rays from a source on the logging sonde. When appropriate corrections have been made, the response can be related to the electron density of atoms in the formation, which is in turn a function of the bulk density of the formation. Knowing the density of the solid rocks allows the porosity to be calculated.

The density and neutron logs are often interpreted alongside one another, as this allows more subtle variations in porosity, lithology and formation fluid composition to be detected than would be possible with either log in isolation.

The sonic or acoustic log measures the time taken for a sonic pulse (a sound wave) to pass through a given thickness of formation. This is known as the 'interval transit time', and is commonly measured in microseconds per foot. It is therefore the inverse of sonic velocity. The interval transit time depends on the nature of the lithology and the porosity. It can thus be used to indicate broad changes in porosity, and is also valuable for lithological correlation between wells. The sonic velocity is also used by geophysicists in converting the time taken for a seismic pulse to travel to and from a given geological horizon, to the true depth of that horizon.

This brief summary of a few logging tools and some of their applications is necessarily simplistic, and a thorough understanding of electric logs requires experience and training. No geologist, however, can afford to ignore wireline logs if they are available. They provide detailed information on the well, often over its entire depth. They allow cored intervals to be placed in their stratigraphic context, and by careful calibration over cored intervals can be used to reconstruct detailed lithological sequences within and beyond cored sequences. They are also indispensable as a means of correlation between wells. Without cores for calibration, electric log interpretation is a rather ambiguous and uncertain process. Conversely, without electric logs, cores that do not cover a significant proportion of the total well depth are of limited value. They are, in effect, spot samples, with an uncertain relationship to the total succession drilled.

When comparing cores with electric wireline logs, it is important to consider the resolution of the wireline tool. A resistivity tool, for example, essentially measures the resistivity between two points, which could be, say, 1 m apart. The resistivity is therefore summed over an interval rather than measured at a point. The resolution of this tool would be of the order of 1 m, which means that, in the absence of additional information, the thickness of a high-resistivity coal bed, for example, could be measured to an accuracy of about 1 m. In the case of resistivity logging tools, some devices have much higher resolutions—of the order of several centimetres or less. The majority of wireline logging tools of all types in common use, however, have resolutions that are considerably lower than this, so the exact location of lithological and other boundaries cannot be precisely determined.

It is important to understand the difference between driller's and log depths (the 'core-to-log shift'). All depths in a well are measured from a certain surface datum, such as the kelly bushing (Section 2.2). The driller measures the total depth of the well as equivalent to the total length of drill string below that datum, when the bit is resting on the bottom. The logging engineer measures the length of cable that has been paid out, and may apply some correction to allow for stretching.

It is not surprising that the two depth measurements do not necessarily agree at the bottom of a 3-km-deep hole. Discrepancies of 10 m or more are not uncommon, although agreement is usually better than this nowadays. By convention, all depths are usually converted eventually to log depths by comparing the core with one or more of the logs (Section 6.7.3). The gamma-ray log is usually the simplest, especially if core gamma measurements have been made (Section 5.2.1). However, this can be done only when all the data are to hand. The depths marked on the cores are thus driller's depth (sometimes called, rather ambiguously, 'measured depth'; see also Section 4.3.2). Thus all analyses carried out on the core, including core logging, are usually referred to driller's depth.

2.9 Sample damage due to coring

The penetration of a well through a rock formation and the recovery of part of that formation in a core barrel are, from the rock's viewpoint, very stressful operations. In addition to the forces involved, all sorts of drilling fluids and associated solids, which might be quite incompatible with the natural formation waters, are introduced. In these circumstances some damage to the cores is only to be expected.

The amount and type of damage will depend on numerous factors, such as the drilling technique employed, the nature of the formation penetrated, the depth, and the drilling fluids used. The responsibility for ensuring that the core brought to the surface is in the best possible condition is usually that of the driller rather than the geologist (although the geologist, at the risk of incurring his wrath, should keep a close eye on the driller and warn him in advance of any geological factors that might affect sample quality). The geologist must, however, be aware of the types of damage that may occur, so that the damage is recognized and accounted for correctly during sample description.

Even before the coring bit reaches any given point in the formation, damage can occur at that point. The rock may suffer compaction, for example, due to the weight of the approaching drill string. Compaction is likely to be significant only in shallow wells, however, and at any moderate depth the opposite can happen. The removal of overburden from the formation may cause it to expand, or even to fracture, as a result of the stress release. Vibration and rotational stresses propagated from the bit may also damage some lithologies, especially unconsolidated ones.

In addition, the drilling mud, which will usually be at a higher pressure than the formation fluid, will penetrate porous lithologies ahead of the bit (Section 2.4). The mud filtrate will easily flush through a permeable lithology for tens of centimetres, so that fluids recovered in a core may bear little comparison to the formation fluid. This will be of particular concern to the hydrogeologist or petroleum geologist, for whom the contained fluid is ultimately of greater concern than the rock itself.

Furthermore, in some cases the introduced fluids can alter the rock mineralogy. The most obvious example is where certain water-sensitive clays (notably smectites) expand, so that they lose their original structure. This also tends to make them soften, so that they are liable to smear out or disperse in the drilling fluid. Fortunately, this problem is well recognized by most drillers, and various conditioners can be added to the mud to prevent it. It can still, however, be a serious problem, and one of which the geologist should be aware.

Despite being permeable to mud filtrate, most rocks are effective filters of mud solids. In practice it is rare even for fine-grained components of drilling mud to penetrate a consolidated core in significant amounts to a depth greater than, say, 0.5 cm. Of course, such invasion would not greatly affect core descriptions, but could have a significant impact on, for example, mineralogical analyses and permeability determinations. The possibility of deeper invasion of a permeable core by mud solids, although uncommon, must always be borne in mind.

In Section 2.3, the possibility of thermal metamorphosis of cuttings by the drilling bit was highlighted. Because of the relatively slow and low-pressure operation of the coring bit, this is not a common problem with core material.

However, it is still safest to assume that particularly sensitive material, such as organic matter and hydrocarbons, may have been thermally altered around the margins of a core.

Several other types of core damage occur as the core is being cut and is rising up into the core barrel. The section of core that has just been cut may jam within the bit and became detached from the formation immediately below. The detached core will continue to turn within the rotating bit and the grinding action between the core and the formation about to be cored will generate concentric markings on the interface between the two. Fortunately, although this may cause headaches for the driller, and may be a nuisance when cutting orientated core, it rarely poses serious problems for the geologist. As drilling continues, the core section caught within the bit is generally pushed out into the barrel as the next section of the formation is cored.

Although the core barrel is designed to collect the core with as little disturbance as possible, some mechanical damage as the core enters the barrel is usually to be expected. All rocks have natural zones of weakness, such as bedding planes and some mineralized zones, as well as natural fractures. Core will usually be broken along such features. However great the core diameter, or long the core barrel, unbroken lengths of core greater than, say, 50 cm are uncommon. The stresses involved in rotary coring should not be sufficient to break a well-consolidated lithology with no significant planes of weakness, although poor coring technique or misfortune can result in this. The use of scribing knives (Section 2.6.1) on a relatively brittle rock, such as a silica-cemented sandstone, can result in the propagation of fractures between the knives, resulting in longitudinal splitting of the core. In claystones and mud rocks, coring-induced fracturing within the core accompanied by slight relative movement along the fracture surface sometimes results in the formation of shiny surfaces covered with parallel striae. These are presumably aligned in the direction of maximum strain. They look superficially like the 'slickensides' sometimes formed by natural fault movements, and it is important not to confuse the two phenomena. Because the artificially induced variety is the result of torque within the core barrel, the surfaces tend to have a curved or slightly helical geometry, which can distinguish them from the more commonly planar geometry of true slickensides.

As a core enters the core barrel, it is sometimes marked by the steel prongs of the core catcher, forming a number of longitudinal grooves, which can give a core the appearance of a Greek column! This is usually only a surface effect on relatively soft cores, and rarely causes serious damage.

The inner core barrel is designed to fit the core with only a narrow gap or annulus between the two. Unfortunately, this can generate considerable resistance to the passage into the barrel of 'sticky' lithologies such as wet clay. Considerable drag around the surface of the core may result.

When the core barrel first reaches the bottom of the hole in preparation for coring, it is common to add short lengths to the drill string at the surface, in order that the kelly may be located as high as possible within the kelly bushing. The maximum amount of core may then be cut before a new length of drill pipe needs to be added. To add further lengths, the kelly, with attached drill string and core barrel, needs to be lifted. This breaks the core, which may become jammed in the core catcher so that coring cannot recommence once the connection has been made. Even if coring does proceed satisfactorily, the drilling break can cause damage and disordering similar to that which may occur at the base of one core and the top of the next (Section 4.3.18). The use of a top-drive rig (Section 2.2) removes the requirement to lift the drill string when making connections, and thus prevents this type of damage.

Some forms of core damage occur once coring has been completed, as the barrel is lifted to the surface. As the drill string is initially raised to take pressure off the bit, the prongs of the core catcher are activated. The aim is to grip the base of the core, so that it is broken from the formation. The success of this operation depends largely on the nature of the lithology. Sometimes the bottom section of the core slips out and is lost; at other times, despite slipping, it is retained, but may be deeply marked by the core catcher prongs, resulting once more in a 'Greek column' effect.

In deep wells the core has been subjected to considerable overburden pressure over geological time. The sudden release of confining stress as it is raised to the surface can result in spontaneous fracturing. At the same time, any gases present in the core (either in the gas phase, or dissolved in pore waters or hydrocarbons) may expand rapidly. This expansion, not dissimilar to the effect of removing the cork from a champagne bottle, can push formation fluids rapidly through the pore structure, resulting in the mobilization of authigenic clays and other delicate mineral phases, and thus altering the microscopic rock texture. In extreme cases the gas expansion can cause a virtual explosion or liquefaction of poorly consolidated rock. These effects are minimized by reducing the speed at which the core barrel is raised, and fortunately major damage to the rock fabric, although potentially dramatic, is not common.

This section has considered damage that may occur to a core as a result of the coring process. With modern drilling techniques, good-quality core can usually be recovered from the most difficult of formations, even at considerable depth (although economic considerations often require the geologist to work on core that is not as good as might be obtained). At least of equal importance, however, to how a core is obtained, is how it is handled, both at the well site and subsequently. This question of core handling is considered in the next chapter.

3 Core handling

3.1 Care at all times!

The choice of well location, and the decision to core at a certain depth, may well have been made by a geologist. But the job is then usually handed over to the driller. Although the wellsite geologist often will (and usually should) monitor drilling progress and advise the driller during the cutting of a first core, this is generally a rather slack time. If the geologist present has staked his reputation on what the core will reveal, it can also be an extremely anxious time! Then suddenly the core barrel emerges from the hole, and there is intense activity while it is retrieved from the barrel, laid out, cleaned, measured, labelled, logged, and packaged for transport to the laboratory. These can be among the most exciting moments in a geologist's life. But in that excitement it is vital that the greatest care is taken with the core: that it is not disordered, no pieces are lost, and it is handled carefully and correctly. This requires a methodical approach.

This chapter describes many of the procedures that should be followed when handling core. Precisely how any core is handled will depend on its nature, what it will be used for, how it will be transported, and the nature of the planned analytical programme. Some of the procedures suggested here will not be appropriate for all purposes, whereas some analytical techniques will demand core handling methods other than those described. It is important to consider well in advance the purposes for which a particular core is being cut, and then plan a core-handling programme accordingly.

Sometimes the core will reveal something unexpected, and require certain analyses that were not anticipated (it would not be the first time that a supposed water well has found oil, or an oil well has found coal). It is therefore best to cater for all reasonably plausible eventualities in the design of a core-handling programme.

Possibly the main cause of disordering of the core after recovery is the innocent individual who picks up a piece for inspection, and then forgets either where it came from, or which way up it was. This is done very easily, even by those who should know better. It is good practice to label each piece of core with a way-up indicator (Section 3.2.4), and moreover, when removing a piece, even if only for a moment, to mark the gap. The intention may be to hold the core segment for a few seconds only, but it is easy to be disturbed by a telephone, or by someone else arriving on the scene. The geologist should get into the habit of marking the spot from which core has been removed for closer inspection, with pencil, ruler or similar item; he will be glad he did so.

Core is nice stuff. A length of core containing a colourful mineralized zone, or bleeding oil, is particularly attractive, and looks good on anyone's desk. Many good cores have been damaged by a drilling crew member, cook, or even the geologist, telling himself that a little bit 'won't be missed'. Trained personnel will be aware of the value of maintaining the integrity of the core. Even so, it is best to limit access to those directly involved with the coring programme at any time, and not to leave the core laid out and unattended if this is avoidable.

3.2 Core handling at the wellsite

In general, the inspection and testing of core occurs in two main phases. First, it is retrieved at the wellsite. At this stage it is rapidly undergoing numerous physical and chemical changes as a result of its removal from the possibly high-pressure and fluid-saturated subsurface to the low-pressure, dry, surface environment.

Although these changes will not affect the basic lithology and some other rock properties, they will certainly affect other characteristics, such as the pore-fluid chemistry, and possibly the rock strength and clay mineral morphology. Thus it is often important either to test the cores at the wellsite, or to preserve them so as to arrest the process of alteration, prior to later testing. It is also worthwhile to make a core log at the wellsite, even if this is to be repeated in greater detail later. This may be of immediate value in deciding whether or not to take further core or continue drilling, especially if the decision is going to be made by, or in discussion with, others away from the rig-site to whom the log may be sent. It will also serve as a precaution lest the core is lost or damaged subsequently while in transit.

It is often not possible to undertake all the necessary analyses on core at the wellsite, so it is moved to a laboratory for further work. The laboratory may be a trailer 50 metres away, or a large complex in company HQ on the other side of the world. In either case, it is a location provided with at least basic facilities where the core can be laid out and studied, away from the bustle and distraction of the drilling floor.

Wellsite procedures for core handling tend to follow similar basic principles no matter what the purpose of coring. These procedures will be considered here. The handling of core away from the wellsite is much more dependent on the specific aims of the coring programme, and will be discussed later (Section 3.3).

3.2.1 Procedure during coring

As mentioned above, during the cutting of the first of a series of cores the geologist's workload is relatively light. Once the first core has surfaced there is generally a race against the clock to get it laid out, described, sampled and packed (or whatever) before the next core arrives. The time taken awaiting the first core barrel should not be idly spent, however, as there are several important operations to carry out.

First, a final check needs to be made that all the tools and equipment required are to hand. Stocks of consumable items will of course have been checked earlier, while there was time to amend any omission. However, the geologist should now ensure that they are laid out, and immediately available. A checklist of some such items is given in Appendix 2.

Boxes or troughs must be set out on the drill floor (but not in the way of the drilling crew). These are for initial laying out of the core after removal from the core barrel. They may be stout wooden boxes, perhaps 1 m long and of sufficient width and height to accommodate the core without it rolling around. The boxes used for onward transportation from the wellsite may be used (Section 3.2.7), but it is often more convenient to have separate boxes designated for core-catching, so that core may be moved from one to the other after reordering, cleaning and so on have been completed. If there is insufficient space on the drilling floor for all the boxes to be laid out at once, they must be stacked in such a way that they can rapidly be retrieved in the order required (Fig. 3.1).

Alternatively, lengths of rigid plastic or metal guttering, or angle iron fastened to a bench or trolley, may be used. These core-catching containers should be numbered consecutively from 1 up, and marked with arrows to indicate which end will house the top of the core (the arrows conventionally point upwards). Alternatively, they may simply be labelled 'top' and 'bottom'.

Since there is unlikely to be time to fit the core pieces closely together at first, the total length of these boxes or troughs should be greater than the core barrel length. It should be ensured that they are completely free of fragments of previous cores, or other debris.

Any materials that will be required for preserving and packaging the core should be prepared. Plastic sleeving (if used) can be cut to appropriate lengths (Section 3.2.6). If sections of core are to be preserved in wax or similar material, the wax bath should be prepared and switched on (Section 3.2.6).

Core boxes for packaging and transport of the core from the wellsite may be prepared (Section 3.2.7). Although detailed labelling will not yet be possible, they may be marked with the well name or number, together with 'top' and 'bottom'.

In addition to this preparation for the arrival of the core, the geologist must monitor the coring progress and be prepared to discuss any operational difficulties with the driller. The cuttings produced by a coring bit are usually finer than tri-cone bit cuttings, but they should still be examined regularly in case of poor core recovery. The rate of penetration of the drill is also a very valuable indicator

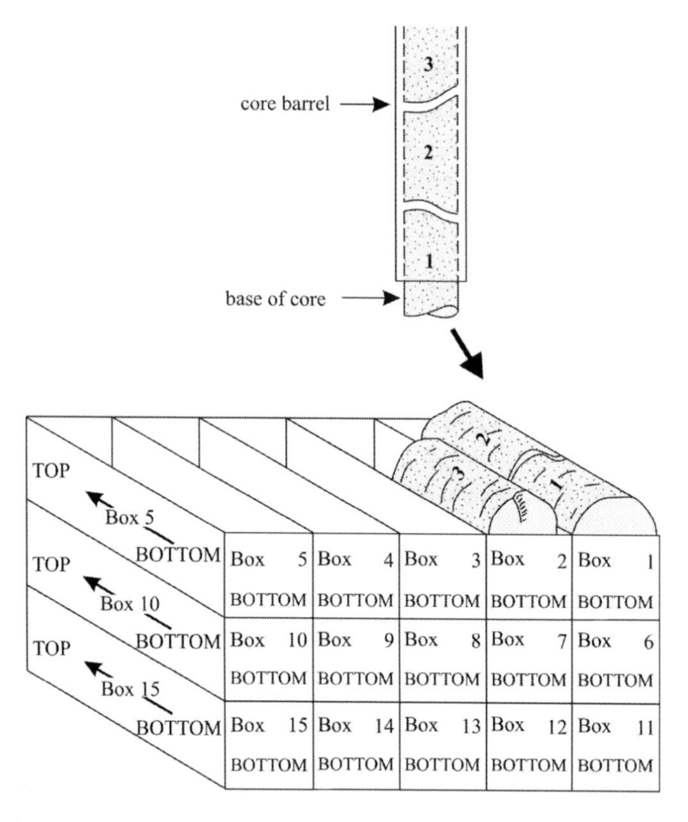

Fig. 3.1 *Core-catching boxes need to be laid out or stacked at the wellsite in order, and properly labelled, so that they are readily accessible for receiving core as it emerges from the barrel.*

of lithologies, and the time taken to penetrate each 25 cm (say) of formation may usefully be recorded.

If there is a chance of any mineralogical or chemical analyses being undertaken on the core, the mud composition (including any additives) should be recorded, and a sample of mud taken at the shale shaker for dispatch to the laboratory.

3.2.2 Core retrieval

On retrieval from the well, relatively short core barrels may be laid down on a bench, or drill floor. After detaching the bit and core catcher, and removing any core from within them, it may be possible to push the core out of the barrel using a rod or plunger. Mechanical pistons are available for this task. Where possible, it is best for the core to be extruded in the same direction as it entered the barrel (i.e. pushed up through the top of the barrel), so that stresses on the core are not reversed, causing unnecessary internal movement. As the core emerges, it must be placed carefully and in order in the core-catching boxes set out earlier, starting with Box 1: this is of course obvious, although in the heat of the moment what is obvious is not always what actually occurs!

If the core has been taken from the top of the barrel, the first piece to emerge is from the highest part of the cored interval and must be placed at the 'top' of the first box. If taken from the bottom, then the reverse is true. Compressed air or water at high pressure may be used to force the core from the barrel in case of difficulties, although water should not be used if water-sensitive minerals may be present in the core, and neither air nor water should be used if fluid measurements are subsequently to be undertaken. Although this method is usually quite effective, great care needs to be taken when using it, as core tends to emerge slowly at first, and then shoots out uncontrollably. It is not uncommon to see core pieces fly out over the horizon or into the sea. The geologist should not risk injury by standing in front of the core barrel when compressed air or water is being used.

With many types of core barrel, the sign that the entire core has been removed is the appearance of a metal slug known as the 'rabbit'.

With longer core barrels it is often neither practicable nor possible for them to be laid on the drilling floor. There may simply be no room—especially on an offshore platform—and in any case the most efficient method of removing a long heavy core is generally by gravity. The inner core barrel is therefore hung in the derrick, and slowly raised by the driller a little above the drilling floor. As the core slides out, the pieces are removed in sequence and placed the correct way up and as tightly as possible into the core-catching boxes or troughs. Other than with small-diameter core, one of the drilling crew will control the movement of core from the barrel using core grips. To avoid injury it is important that no one puts their hands underneath the core barrel. Excessive lengths of core may be broken, as cleanly as possible, with a hammer. Any disaggregated or rubbly sections must be collected as carefully as possible, and excess cleared away so that it is not mixed with subsequent sections of core.

Pieces of core will frequently slide out of the barrel suddenly and forcefully. In addition to the danger of injury, this is also a major cause of core breakage and disordering. The geologist must take great care in both respects.

Sometimes the core will jam, and not come out of the barrel easily. This is usually overcome by a few blows on the barrel with a light sledgehammer, although care must be taken to avoid damaging the barrel. On the rare occasions when the core remains stuck, the barrel may be laid down horizontally (if there is room) and the core expelled with compressed air or high-pressure water, as described above. Great care needs to be taken to avoid loss of or damage to the core, or injury to person, since this operation can bear similarities to the firing of cannon! As a last resort, the core barrel will need to be cut open, which may require its dispatch to the laboratory.

The core will have been subjected to stress relief and partial draining while being brought out of the hole, but it is upon removal from the core barrel that it will begin to dry, and can deteriorate rapidly. The geologist will not have time in these first few minutes to complete a written description of the core. However, a careful mental note should be made of features of the core (bleeding oil, bubbling gas or smell) that might no longer be obvious once the core is laid out. A written note of such features should be made at the first opportunity.

These guidelines for dealing with cores do not, of course, apply to those retrieved in aluminium, fibreglass or rubber sleeves. These are retrieved intact from the core barrel and transported direct to the laboratory, perhaps after cutting into manageable lengths and sampling (Section 3.2.5).

The driller will not want the geologist to remain with the core on the drilling floor for long, as this would interfere with the continuing drilling procedures, and possibly waste valuable rig time. The core-catching boxes should therefore be removed without delay to an area where they may safely be laid out for cleaning and inspection.

3.2.3 Cleaning the core

The degree of cleaning needed by a core depends largely on the drilling fluids used, and on the nature of the formation. A core through a well-consolidated clean sandstone or shale drilled with water will probably emerge from the core barrel in a spotless condition.

The use of drilling muds, or the presence of soft clays in a succession, can alternatively produce a core completely caked with mud. As much of this as possible should be removed quickly, not only because the degree of contamination by the mud may increase with time, but also because the core is much more difficult to clean when dry.

Cleaning should be undertaken with a fluid compatible with that used to cut the core. Water, for example, should not be used to clean an evaporite core drilled with salt-saturated mud; nor should it be used with an oil-base mud. Care should also be taken if there is the possibility of water-sensitive clays, such as smectites, in the core.

In any event, the cores should not be hosed down, as this is liable to cause disordering of the core, and will also jeopardize any fluid saturation measurements that may subsequently be made. At sea, water from a hose would almost certainly be seawater, which may precipitate salt in the core on drying. For the same reason, cores should not normally be immersed in fluid. The safest method is to wipe the cores with rags moistened with an appropriate fluid, such as water or the base oil. Thickly caked mud may best be removed simply by scraping with a blunt knife, and this is also the best technique with highly friable or permeable cores, or those containing evaporites and drilled with saturated water-base muds.

3.2.4 Measuring and labelling

At this stage the cores will normally still be in the core-catching boxes, troughs or similar. This may be convenient, but if they are to be transferred at some stage to other boxes for transport or permanent storage this may be undertaken now (Section 3.2.7).

Alternatively, if the cores are being kept in deep core-catching boxes, it may be helpful to lay them out at this stage on lengths of angle iron or guttering, where they will be more clearly visible and easier to pick up. This is largely a matter of convenience, but the more often the cores are moved, the more opportunity there is for disruption, so great care needs to be taken.

If there is sufficient room, and the core is in individual short boxes, these are usually best laid end to end in order, rather than lined up side by side. Not only is the core then more easily accessible, but also it will be displayed in its proper geological sequence, and thus be more meaningful to the geologist. In practice, space constraints often require short lengths of core to be laid out side by side, and indeed many core boxes consist of trays already divided into a number of parallel compartments.

The next step is to refit the core, prior to measuring and labelling. As with cleaning of the core, this needs to be carried out quickly if subsequent steps are to be taken to prevent further drying and deterioration. Starting at the top of the core, the pieces must be fitted together as closely as possible, so as to form a continuous cylinder. Many adjacent pieces of core will fit neatly together. When this is not the case, a piece may have become disordered, and attempts may be made to fit the core piece into other gaps nearby, or upside down. If there is no obvious fit, however, the core piece must be returned to its former position. Some geological knowledge will be of use here, such as the recognition of unequivocal way-up structures in the core. The geologist must be certain, though, that the geological succession cored was not inverted!

Furthermore, the geologist must be careful not to use preconceived models to reconstruct the core. If, for example, a section of seat earth has become disordered in a core containing a coal bed, it must not be assumed that the seat earth came from below the coal without good supporting evidence. Such an assumption might destroy vital evidence for the washout of a coal seam. It is safest to leave the disordered seat earth where it is, but to make a note in the core description (Section 4.3.18) that this piece of core may be out of place.

Any rubble and disaggregated core should be placed into spaces between core pieces that do not fit neatly together. It is best to put such rubble into sample bags, which will subsequently be marked with the appropriate depths. If the whole core, or long sections of the core, is disaggregated, it should be placed into a number of bags, maintaining them in order as much as possible.

Once the whole core has been fitted in this way (Fig. 3.2), the total length recovered must be measured. Care needs to be taken to make proper allowance

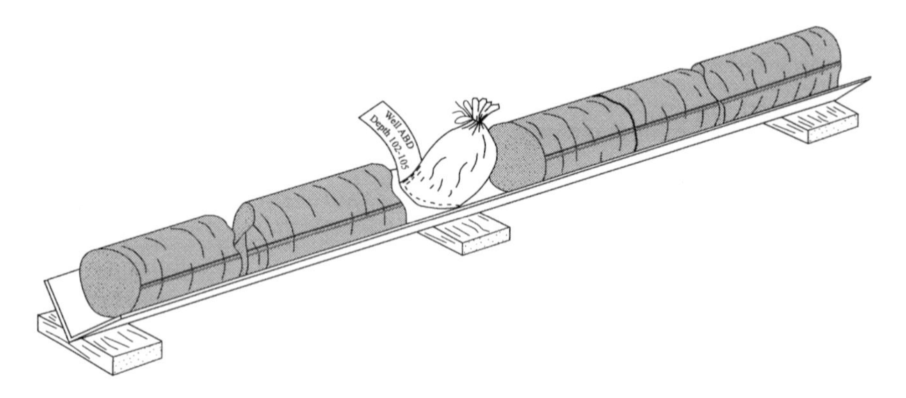

Fig. 3.2 *Core pieces fitted together on a length of angle iron. A rubbly section is contained in the bag. The parallel lines marked on the core, which would be in contrasting colours, indicate the way up.*

when moving from one core box to the next, where the depth of the lowest point of an unevenly broken piece of core at the base of one box will not coincide with the depth of the highest point of the underlying core. The measured length must be compared with the driller's figure for the thickness of the cored interval (which will be less than or equal to the length of the core barrel).

If the measured length exceeds the cored thickness, an attempt should be made to achieve a closer fit of the core. A better fit can sometimes be achieved by reorientating individual core pieces: the proper orientation is of course much more easily achieved with scribed core (Section 2.6.1). It is also common to allow too much space for rubbly and disaggregated core; unless some is known to have been lost, it should be squeezed up so that, by a visual estimate, its total volume would fit into a cylinder equal to the diameter of the core.

If the measured length of the core falls short of the cored thickness according to the driller, it is assumed in the absence of evidence to the contrary that the deficiency has been lost from the bottom. There are methods of testing this assumption that may be applied later (Section 4.3.18). If, as often happens, the lowest section of core has simply slipped out of the core barrel as it is lifted up the well, it may be retrieved, usually in a rubbled state, at the top of any immediately subsequent core (which will obviously complicate depth measurements in both cored sections). Core is also sometimes found to be missing as a result of 'milling' – where the drill bit has ground away the rock entirely instead of cutting whole pieces of rock. The lost core will usually have been carried away in the drilling fluid like conventional cuttings.

The ratio of the length of core material recovered to that cut (according to the driller) in each core, expressed as a percentage, is defined as the 'total core recovery'. Engineering geologists also define 'solid core recovery' as the proportion of core recovered that is composed of individual lengths of full-diameter core rather than fragmented material. Solid recovery takes longer to measure than total recovery, but is more significant in engineering terms. A note should be taken of the total core recovery percentage, together with the solid recovery if appropriate.

Once the core has been reconstructed to the geologist's satisfaction, it must be marked or labelled to indicate depth and orientation. The markings are normally most easily made on the core with a permanent black felt marker or wax crayon (if the latter, it should not be soluble in oil). If the core is too soft or friable to be marked permanently in such a way, it should be transferred to the box that will be used for transport and long-term storage (Section 3.2.7), and the markings made on this.

There are various methods of marking the core. A good way to mark depth is to make a line, starting at the top, around the circumference of the core at regular intervals (every 25 cm, or 1 foot, say). If the core is of sufficient diameter, the relevant depth (or failing that, the last couple of digits of the depth) should be written above the line.

To mark the orientation, one may draw a single line along the length of the core. The line should continue uninterruptedly across any breaks in the core (Fig. 3.3), and should then be annotated along its length with arrowheads, pointing in the direction of decreasing depth, ensuring that each core piece has at least one arrowhead. Fragments of core that are not transected by the line along the length of the core, but for which a 'way up' is known, may be marked with a separate small arrow.

An alternative method for marking orientation, which is a standard method in the hydrocarbon industry, is to draw two parallel continuous lines, close together, along the length of the core (Fig. 3.2). They should be of different colours, such as red and yellow, with the red line on the right when the core is orientated the correct way up. This is quicker than drawing arrows, and remains intelligible if the core is further fragmented, whereas arrows are meaningful only if the arrowhead remains in place. Unfortunately, there is no universal convention as to which colours should be used, and on which side each colour should be.

If the core has been scribed, it is helpful to draw the way-up arrow, or one of the coloured lines, along the primary scribe line.

3.2.5 Sampling and reporting

Once the core has been measured and labelled, it is ready for sampling and/or description, and for any wellsite testing to be undertaken. In some instances the core will be fully described and tested at the wellsite. It might then be disposed of, although, considering the expense of obtaining core, it is usually better to retain it while the project in question is continuing (which is often a contract condition for commercial work), with consideration given to long-term storage (Chapter 7).

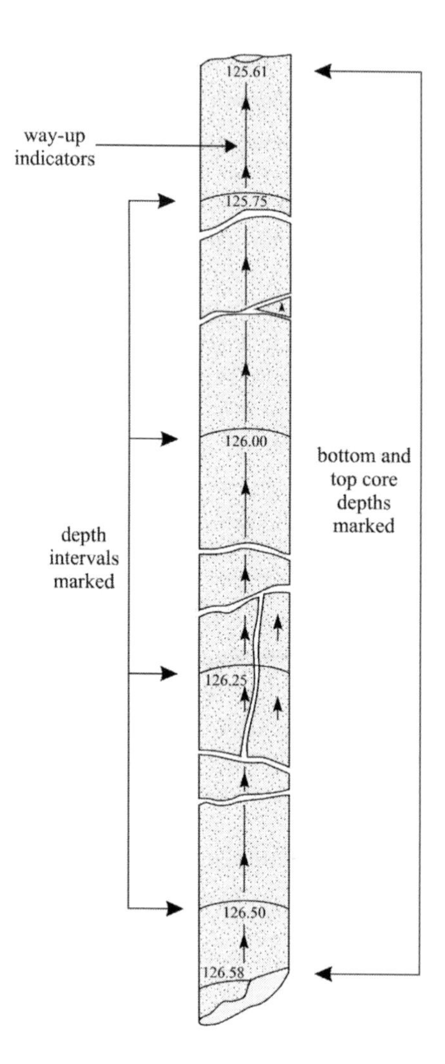

Fig. 3.3 *Depths marked at regular intervals along a reconstructed length of core, and at each end. Way up is indicated by arrows.*

However, even if a full description and analytical programme is to be undertaken later in the laboratory, it is good practice to describe the core and take samples at the wellsite. Not only does this allow any interpretative work to commence at once, it also guards against total disaster should the core be lost or damaged in transit (an event that is not altogether uncommon).

Of course, the description and selected samples should be dispatched separately from the core, a good arrangement being for the geologist to carry them back to base once the assignment has been completed.

Methods of describing core are detailed in the next chapter, and will vary considerably depending on the reasons for the drilling of the well. So far as possible, tests should be non-destructive, and every effort should be made to prevent the core from becoming disrupted (Section 3.1).

Once the description has been completed, sampling should be carried out. Again, the size and frequency of samples will depend on the total length and diameter of the core, and on the types of analysis that may be undertaken. Generally, the removal of samples should make little impression on the core as a whole. This is particularly important if the core is to be subject to analyses in the laboratory that will be jeopardized by the removal of significant quantities of core (such as core-gamma measurements, Section 5.2.1). For the same reason, there should be sufficient documentation to ensure that the core piece can be reinstated in its proper position within the core at a future date. This can be ensured by slipping a note, written with waterproof ink on thin card or similar material, into the core-box in the appropriate gap. The note should record sample depth, reason for sampling, collector's name, and date.

If the sample is to be sent to an outside contractor, or to one of many laboratories or sites within an organization, this should also be documented. It may be easiest to use a small preprinted card (Fig. 3.4). An identical sample card could accompany the sample.

The sample itself may be marked with a way-up indicator, and placed in a bag labelled with well number and depth. Resealable plastic bags with white panels for labelling are simple to use and relatively cheap, but they are sometimes prone to coming open in transit, and rock samples with sharp edges can wear holes through them. Bags of linen (or similar tough fabric) with a drawstring and sewn-on reinforced paper label are to be preferred.

In addition to labelling the sample bags and core, two separate lists of sample depths should be made, one to be shipped with the samples, and the other to be retained by the geologist.

If the core has been recovered in a fibreglass core barrel liner (or similar), it will only be possible to take a sample at each end. However, if the core is cut into

CORE SAMPLE

Well: ....

Depth: ....

Sampled for: ....

Collected by: .... Date: ....

Despatched to: ....

Fig. 3.4 *Example of a core sample card. The card can be completed in duplicate: one copy to accompany the sample, and the other to be placed at the appropriate position in the core box.*

shorter lengths at the wellsite for onward transportation, it will be possible to take a larger number of samples.

Some laboratory analytical techniques require samples to have been specially preserved at the wellsite. This is discussed in the next section.

3.2.6 Core and sample preservation

The drying out and deterioration of a core are not always significant problems, and indeed detailed core descriptions are often best undertaken once the core has been cleaned and allowed to dry. However, there are several analyses that are valid only while the core remains in its original fluid-saturated condition.

Because it is not usually practicable to carry out such tests at the wellsite, the cores, or subsamples from the cores, need to be preserved in some way so that they do not deteriorate further on their way to the laboratory. Similarly, as it is often the weakest or most easily deformed sections of core that are of greatest significance in geotechnical studies, it is sometimes desirable to adopt special preservation techniques to enable these lithologies to survive intact as far as the laboratory.

There are a various preservation methods that may be adopted at the wellsite, and some of the most common are described below (Fig. 3.5). As has been emphasised before, it is of the utmost importance to work quickly, so that core degradation can be arrested (or at least slowed) as early as possible. In the time taken to cut the core and retrieve it from the well, a significant amount of alteration will already have occurred. The most damaging process that most cores undergo is being cut from the formation, lifted up the well-bore, and removed from the barrel. The extra short time taken in measuring, labelling and briefly describing the core should thus not unduly increase the degree of alteration, but nevertheless, the more quickly it is completed, the better will be the quality of the preserved sample.

Possibly the simplest method of preservation is to enclose the core in clear polythene 'lay-flat' sleeving. Similar but tougher (and more expensive) proprietary sleeving made (for example) of metallized plastic is also available.

Lengths of sleeving of diameter slightly greater than the core are cut, usually a little longer than the core-box length, and the core inserted. The ends may be sealed either by folding over several times and stapling, or by heat-sealing. The latter method forms a more effective seal, but some practice with the heat-sealer is sometimes required to create a weld that holds together, but which does not melt a hole in the plastic. (Some of the proprietary materials are better in this respect.) Another disadvantage of the heat-sealing method is that if it is necessary to reopen the sleeve, it has to be cut, and cannot then effectively be resealed. It is also much easier to seal a plastic sleeve containing a length of heavy core with a staple gun than a heat-sealing iron. At the wellsite, the use of a stapler is generally the best option.

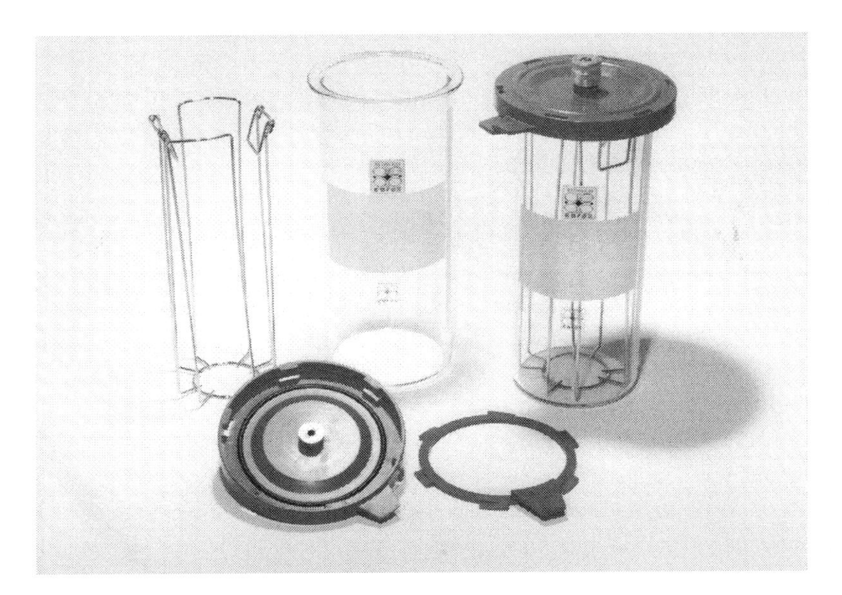

Fig. 3.5 *Common core preservation methods. The simplest method is to wrap the core up in, for example, aluminium foil. A double layer, perhaps with cling film, is more effective. Enclosing the wrapped core in lay-flat polythene sleeving, with heat-sealed (or tightly rolled and stapled) ends will provide further protection. At the front is a sample that has been sealed in wax. The lower picture illustrates the sealed glass jar (Photo courtesy of Corex Services Ltd.)*

Even with care, it is difficult to seal core in a polythene sleeve without leaving significant airspace inside. This will cause the core to continue drying to some extent, and possibly to move around, with the danger of breakage and disorganization. This can be minimized by first wrapping the core with plastic clingfilm (Saran wrap) and/or aluminium foil. The combination of clingfilm, foil, and outer polythene sleeve forms quite an effective seal, at least over a few days.

No preservation method is suitable for all eventualities, and it should be noted that clingfilm, although excellent for preservation in many respects, exudes volatile 'plasticizers', which may affect organic geochemical analyses (see below), and also the wettability of core. Wettability is a parameter that is sometimes measured, because it influences the production of hydrocarbons from a reservoir. If wettability analyses are planned, the clingfilm should not be allowed to come into contact with the core.

For more complete and longer-lasting sealing, lengths of core may be dipped in wax, or one of various types of proprietary resin that melt on gentle heating. This preservation technique is used in the oil industry on samples to be used for whole core analysis of, for example, capillary pressures and flow performance. Such tests are grouped under the heading of 'Special core analyses' (Section 5.2.6), so the waxed core pieces are known as SCAL samples. SCAL samples are first wrapped in clingfilm and foil to improve sealing, and to prevent contamination of the core by wax.

The core is then labelled, suspended from a string or wire, and dipped in a bath of molten wax so that, on cooling, the wax forms an impermeable covering to the core. If string or other porous material has been used to suspend the core, the ends should subsequently be cut off and themselves dipped in wax, or else the string will form a conduit for the loss of fluids. Wire is sometimes used because it is non-porous and will avoid this problem, but great care needs to be taken, as the tension in the wire when it is used as a handle can easily cause it to cut through the wax. The wax should be translucent, allowing the label underneath to remain legible. The SCAL sample is returned to its correct position and orientation in the core box, so that it will register on the core-gamma log (Section 5.2.1).

The criteria for choosing the location and length of samples for SCAL and other analyses that need sections of whole core will clearly depend on the type of analysis to be undertaken. It is usually important to choose a sample that is representative of the lithology at about that depth. However, it should be remembered that once the sample is preserved it will generally be unavailable for geological description. For this reason, in addition to making a note of the lithology prior to preservation, it is preferable where possible to avoid taking samples that contain diagnostic geological features such as sedimentary structures and bed boundaries, unless these are well represented in the remaining core.

An alternative to the waxed SCAL sample is the sealed glass jar. This is similar to the type of jar with a sealed lid used by cooks for bottling fruit. The whole core is inserted and submerged in a suitable fluid. The top incorporates a gas valve and, after sealing the lid, an inert gas (such as nitrogen) may be pumped in to create a slight positive pressure. The use of these jars is much simpler and more foolproof than waxing. Since they do not need to be heated, as with waxing, there is no danger of fluid boiling off. In addition, the core may be viewed through the glass. However, although the jars are reusable, their cost is high, and the preservation of a number of whole core pieces by this method requires a significant capital outlay. The jars also, of course, require particularly careful handling to avoid breakages, and are transported in custom-built foam-lined boxes.

The preservation of core pieces for organic geochemical analysis poses particular problems. The samples should be kept in an airtight (and preferably air-free) environment (to prevent drying and bacteriological degradation), and must be free of the slightest contamination by organic substances. Neither clingfilm wrapping nor wax sealing fulfils either of these criteria. Glass bottles have been used, but even these usually have metal lids with plastic, rubber or waxed cardboard seals, which are liable to contaminate the sample.

A common solution is to use metal cans with removable metal lids (like large, old-fashioned cocoa tins). The samples are placed in the cans, which are then submerged in salt water or drilling mud to which a bactericide has been added. The lids are secured with metal clips, and the cans stored upside down. It is important that the cans are not allowed to rust.

Another useful preservation method, especially for friable sandstone samples, is freezing. Large mobile freezing units are available into which the boxed core (preferably wrapped in clingfilm and/or aluminium foil) is placed. The water within the core freezes, thus preventing further drying, and effectively turning a friable sandstone into a solid block of ice. This is a satisfactory method of preserving fluids, and will protect the core during the rigours of transportation to the laboratory, during which it might otherwise become completely disaggregated. If a core is not fluid-saturated, it may be necessary to add water or brine to allow freezing. This will, however, limit the value of subsequent fluid analyses. The expansion of ice crystals within the pore system is likely to cause some damage to the microfabric, and, on thawing, the rock may be more friable than before. Analytical results of, say, rock strength and permeability testing may be affected by the core having been frozen. However, it is sometimes better to have some results from a slightly suspect core than no results from a totally disaggregated core.

Experience shows that freezing of core is not as destructive of fabric as may be imagined, and core freezing is a valuable preservation method in certain circumstances.

Core taken in a fibreglass (or similar) core barrel is already, to a certain extent, preserved. At the wellsite it may be cut into manageable lengths, and caps placed on the open ends. It can be transported in this state, or may first be frozen. Freezing will not only protect the core within the fibreglass sleeve during transport, but may also assist with the safe removal of the core once it reaches the laboratory.

Another technique intended to preserve core recovered in plastic or fibreglass sleeving, and particularly for protecting friable or unconsolidated cores, is wellsite resination (Worthington *et al.*, 1987: see Bibliography). This is designed for use with core sleeves in which a narrow annular space remains between their internal diameter and the margin of the core, but it may also be suitable for cores with a very open fabric, such as unconsolidated coarse gravels. One end of the sleeve is capped, and low-viscosity, quick-setting resin is poured into the other end, so that all the spaces within the sleeve are filled. The sleeve must be filled with resin so as to submerge the core completely. This may require a short additional section of sleeving to be taped to the open end, to receive additional resin. Once cured, the core and sleeve are carefully cut open along their length using a diamond saw.

Some of the commonest preservation methods have been described in this section. The number of different ways of preserving core is, however, unlimited. It is up to the geologist or engineer to use common sense to devise appropriate methods depending on the state of the core, and the particular analytical techniques to be used.

3.2.7 Packaging and transport

Between the wellsite and the laboratory, the core will probably need to be transported across rough country or the sea, and may be handled by people who are unaware of its vulnerability. To the untrained, a load of 'rock samples' does not sound like a cargo requiring careful handling. Effective packaging is thus of considerable importance.

Cores are traditionally packaged in wooden boxes, preferably of stout timber resistant to splintering, such as marine ply. If solid wood (rather than plywood) is used, it should be at least 10 mm thick. These boxes are sometimes in the form of wooden trays divided into a number of parallel sections, into which several lengths of core fit snugly. These usually need two people for lifting, which is made easier if a rope handle is provided at each end. Wider-diameter core is packed into individual wooden boxes, each perhaps 1 m long. The tops of these wooden boxes may be fastened on by nails or screws, or with metal straps, ready for transport. Nails should not be used if the core is poorly consolidated, as the necessary hammering may cause damage. Some wooden boxes have hinged lids, and may be fitted with a hasp and staple for securing with a padlock. These do not always hold the lid on firmly, however, and some additional means of fixing the lid during transport would improve security.

Despite the undoubted sturdiness of these wooden boxes, and suitability for handling when palletised by crane or forklift truck, they are usually too heavy for easy hand-carrying, at least over any distance. Wooden boxes are also rather bulky, and are particularly wasteful of space when empty and requiring transport to, or storage at, the wellsite. Because of this, core boxes made of thick cardboard or corrugated plastic are now widely used for transportation and long-term storage. These are supplied flat, and are folded into shape as required.

Cardboard boxes are not really suitable for holding very heavy core, as they tend to sag in the middle, and lose much of their strength when wet. However, the corrugated plastic boxes are excellent for most purposes, and are now widely used for wide-diameter cores.

Ideally, the internal size of the core box should be only fractionally greater than the core diameter. The core should be transferred into the core boxes, taking great care not to cause further disordering. In addition to the well and core numbers, all depth and orientation markings on the core should be marked at the relevant positions on the core box. Any cards noting samples removed should also, of course, be transferred to the boxes. The core at this stage may already be wrapped in clingfilm, foil or similar material, which will help to prevent movement of core in the box, but some additional packing will not go amiss, especially if the core is broken or contains missing sections. Anything soft and voluminous will suffice, such as screwed-up newspapers, straw, polystyrene beads or bubble wrap. However, the best types of packing will not degrade with time, and will be easily removable.

Expanded polystyrene moulds are available that fit tightly into a core box. They comprise two identical halves, top and bottom, between which the core sits in a cylindrical hollow. They are excellent so long as the core diameter is known in advance, and so long as no core pieces (such as waxed samples) will be greater than this diameter. A simpler and quite effective packing material is old rags, which are commercially available, or can be obtained in the closing minutes of some jumble sales, although these become messy when used with wet core. Foam rubber sheeting is also suitable, but this does disintegrate in the long term.

Packaging should ensure not only that core does not move around, but also that gaps in the sequence (due to core loss, breaks between separate cored intervals, or removal of whole-core samples) are faithfully retained. Wooden blocks or dowelling cut to fit neatly into the core boxes are good for this purpose, but tend to be used only on shallow onshore wells where the problems of their bulk do not outweigh their advantages.

Wooden boxes should be fastened with screws or nails, and cardboard or plastic boxes with heavy-duty tape. Further packaging will depend on the total bulk of core, and on the method of onward transport; cognizance should always be taken, however, of the old adage that if anything can go wrong, it probably will. It is not worth taking short cuts with core packaging.

3.2.8 Handling other rock samples

In addition to conventional cores, several other types of rock sample are retrieved from wells. Cuttings samples are of course routinely inspected and collected. In addition to these, there are percussion and other types of sidewall core (Section 2.7), as well as rock samples occasionally retrieved in a 'junk basket' used to clear debris from the bottom of a hole, or pebbles found jammed in a tri-cone bit.

The quality of these different sample types varies considerably, but they are all worth documenting and returning to the laboratory, since they may help with the final interpretation of the well.

Most such samples are best kept in straightforward polythene (or, preferably, linen or other fibre) bags. Percussion sidewall cores are, however, commonly wrapped in foil and/or clingfilm, and stored in small purpose-made glass jars with screwtop metal lids. If the sample is to be used for geochemical analyses, clingfilm should be avoided, and the plastic or waxed cardboard seal from the lid of the jar should be removed.

3.3 Core handling in the laboratory

Core handling at the wellsite inevitably follows certain general procedures whatever the final aim, but once the core reaches the laboratory (or core store, or whatever) anything may happen to it. It is thus not intended to be specific here as to core-handling methods in the laboratory, but rather to suggest a few general guidelines.

3.3.1 Unpacking and laying out

On receiving core in the laboratory, the first priority is of course to check the boxes received against the manifest, and to pursue any discrepancies. The boxes should be opened to ensure that no damage or disordering has occurred during transit.

There may be no further work planned on the core, in which case it can go straight into storage (Chapter 7). Alternatively, it may be intended to undertake a major programme of work. If this is the case, it is often convenient to lay the core out on tables or benches, where it can be more easily viewed, handled and sampled than if it remained in boxes.

Lengths of angle iron or guttering, like that described in Section 3.2.1, are useful for laying out core. Many core laboratories have these permanently set up on benches (Fig. 3.6). It is of the utmost importance that, once laid out, the core

Fig. 3.6 *These lengths of core have been unpacked and laid out at the core analysis laboratory. They are on heavy-duty trolleys, so that they may be viewed, measured and transported around the laboratory with the minimum of disturbance and effort. (Photo courtesy of Geochem Group.)*

is left undisturbed as much as possible. On no account should it be repeatedly laid out and returned to its boxes, as this is almost certain to lead to disordering and damage. For this reason, core is often laid out on heavy-duty trolleys, so that it may be moved from one area of the laboratory to another, and even wheeled into a temporary store, without any need for handling.

There are some cases where the procedure described above is obviously unsatisfactory. Totally disaggregated core is normally best left in the original core boxes. Frozen core will need to be returned as soon as possible to a freezer (if it did not arrive in one). If the frozen core is packed in heavy wooden boxes, perhaps with additional packing inside, it will be quite well insulated. Although this has the advantage that the core will defrost only slowly while out of the freezer, it also means that it will take longer to freeze again once returned (up to several days). In many cases this will not matter. If, however, it is important to start work on the core quickly, or if the core must continually be kept at as low a temperature as possible (for example for fluid sampling), refreezing must take place as quickly as possible. Removing the lids of the core boxes and taking out packing materials may speed up the process sufficiently. Alternatively, if rapid freezing is critical, the core could be immersed in a bath of liquid nitrogen before being returned to the freezer.

3.3.2 Slabbing and sampling

Cores are often slabbed—that is, cut down their length—in order that they may be viewed along a flat surface free of the drilling mud and dirt that usually contaminate their outer surfaces. Although small-diameter core could be cut section by section on a small bench diamond saw, it is more common to use a heavy-duty saw, through the rotating blade of which the core, mounted on a mobile rack, is moved (Fig. 3.7). (Sometimes the core moves while the saw blade remains fixed.)

Core is usually slabbed through its centre, to produce two equal halves. However, it may also be slabbed off-centre so that one of the resulting sections is larger than the other. The chief advantage of this is that relatively large samples or core plugs can still be taken. Sometimes one of the slabbed lengths is slabbed again at right angles to the first cut, so that the core can be viewed in two perpendicular directions.

If the core exhibits some type of directional structure, it may be useful to slab it in a certain direction relative to that structure. Cross-bedded sandstones, for example, are often slabbed in the direction parallel to greatest dip, so that the actual dip visible on the cut surface will not just be some arbitrary value between the maximum dip and zero. It needs to be remembered, though, that structures may be unclear until after slabbing has taken place, so slabbing in a direction defined by some structure in the core, although a worthwhile endeavour, can be rather hit and miss. Cores within sleeves may be X-rayed or scanned in order to determine the orientation of structures prior to slabbing (Section 5.2.3).

In addition, a single section of core may contain beds dipping in several different directions, and a single slabbing plane cannot be parallel to them all.

With orientated core, it is possible to slab in a consistent direction relative to true north, which is a useful option in some circumstances (see for example Section 4.3.11).

Most slabbing saws routinely use water as a coolant. An oil-based conditioner is sometimes added to the water, but this must be avoided if there is any chance that core pieces may subsequently be used for hydrocarbon-fluid or other geochemical analysis. As with any rock-cutting operation, the possibility that the water may dissolve or react with any component of the rock (such as soluble salts or swelling clays) needs to be considered. Dissolution of salts can be avoided by using salt-saturated water as a coolant, but this presents the danger of salt precipitating within the rock, and can also lead to corrosion of the saw. The use of paraffin oil is a better alternative in this situation, but only if subsequent hydrocarbon-fluid analyses are once more discounted.

Frozen core will often need to be slabbed, but this operation must not allow the core to thaw. Water is therefore unsatisfactory as a coolant in this instance. Liquid nitrogen can be used in place of water, although it does tend to get sprayed around the room, so the operator needs to take appropriate precautions.

Although chips of core, or sections of whole core, are used for many types of geological analysis, it is sometimes necessary to drill out regular cylindrical samples. These are used for certain types of analysis (such as porosity and permeability; Section 5.2.5), and are also a convenient method of removing samples from the centre of a core without breaking it into pieces with a hammer, or making several saw cuts. Plugs are cut with a hollow, cylindrical, diamond-tipped bit mounted on a heavy-duty bench drill. As with a slabbing saw (above), water is the most common coolant, but consideration needs to be given to the use of alternatives where appropriate.

Plugging is most easily undertaken on slabbed core, since the core will rest firmly on its flat surface while being cut. Sometimes, however, the whole core must be plugged to obtain a plug of sufficient length. This assumes that plugs are required that are aligned perpendicular to the core length ('horizontal plugs'). The orientation is often unimportant, but for some analyses (such as permeability) it is critical, and plugs may be required that are parallel to the core length ('vertical plugs'), or oriented with respect to a given structure. In this case it may be necessary to saw the core before plugging, and to clamp the core piece securely while cutting the plug to ensure that the correct orientation is obtained.

3.3.3 Resination

The importance of caring for core at all times, so as to minimize damage and disturbance, and to maximize its useful life, has been emphasised throughout this chapter.

Although core is difficult to destroy completely, it can become so broken and disordered as to be rendered useless, and if disordering occurs and is not noticed it can result in serious and expensive mistakes.

Unfortunately, the only way to reduce core damage to an absolute minimum is to ensure that the core remains in its box on the core store shelf. A more realistic method is to mount the core in clear resin so that it cannot be disordered, and so that further breakage or disaggregation will be minimized. Of course, setting the core in resin will contaminate it, and prevent easy access for sampling, so usually only one slab of core is resinated, and a corresponding slab is kept for sampling purposes. The resinated slab is easy to handle, and is sufficient for general descriptive purposes. It also provides a record of the order of the core at the time of resination, which cannot subsequently be disrupted.

Shallow trays of metal or plastic are commonly used as a mould for resination (Fig. 3.8). The tray should be large enough for a length of slabbed core (1 m, say) to fit, flat side down, with several centimetres to spare on all sides. A thin layer of resin is poured into the tray to form a base before the core is added, and is allowed to cure. This basal layer may be opaque to form a background: white is probably the best colour, but subsequent layers must be clear and free of air bubbles. The length of slabbed core is then laid out on this base, taking care to ensure that it is ordered and spaced out as accurately

Fig. 3.8 *Transparent resin is being poured into shallow trays containing half-cut core, with labels giving information on well number, depth and samples taken. (Photo courtesy of Geochem Group.)*

as possible. Any original small gaps in the cored succession should be retained.

Slips of paper or thin card with details of well and core number, depth and scale should be placed around the core, and the opportunity may be taken to indicate important features. A second layer of resin is then poured into the tray, of sufficient depth to hold the core firmly but not cover it (probably about 1 cm, but this is dependent on the nature and thickness of the core). The resin must be poured into the tray, not over the core, and should be of sufficiently low viscosity to flow evenly around the core, and into any cracks between core pieces. If the core is of a porous lithology it may slowly be impregnated by the resin. As this happens, the level of resin in the tray will fall, and may need to be topped up. Care should be taken to ensure that the core that remains above the level of resin in the tray remains free of resin.

Once this second resin layer has cured (Fig. 3.9), a check should be made to ensure that the core is firmly held by it. The slabbing saw is then used to cut off the segment of core that stands proud of the resin surface. The cut should be made at a small but constant height (several millimetres) above the resin surface, and should be as smooth as possible. The offcut will need to be carefully labelled and reboxed. In the UK petroleum industry this offcut is often used to fulfil an obligation to lodge a continuous slice of all core with the appropriate government department. The resinated slabs are now complete, and comprise a continuous length of core held firmly in resin, but with a flat, unresinated top surface. If carefully made, the slabs are in most cases very durable. They can sometimes break under their own weight if not laid flat or carried carefully, but even this will not usually damage the core markedly.

However, unconsolidated sands and soft or crumbly clays may fall out of the resin if it is inadvertently tipped over. The solution is either to take particular care to keep the resin flat, or to apply a thin third coat of resin over the surface of the core. This has the disadvantage that the core is totally enclosed by the resin, but if done carefully the core can still be seen clearly, and is afforded a large measure of protection.

4 Core logging

4.1 Core logging technique

There is a well-known saying amongst the geological community that the best geologist is the one who has seen the most rocks. There is no doubt that the experienced core logger, having seen a wide diversity of cores, has an advantage over the novice. Nonetheless, the secret of creating a good log is to be careful, to be systematic, and to work with no undue haste. Wide geological experience is not a necessity, so long as the logger keeps a clear record of what is observed in the core.

The geologist does not normally need a plethora of equipment while logging. A pencil, or fine-tipped waterproof pen, ruler and paper may be sufficient, together with a clipboard if no other convenient flat surface is available. Field logs may be drawn in a simple lined notebook or on graph paper, although the use of predrafted logging sheets, divided into appropriate columns and with depth markers down the side, can save a considerable amount of time and effort, thus giving a neater (and therefore more precise) result (Section 4.2.2; Fig. 4.1). A hand lens is needed for all but the most superficial examinations. The type with a graticule for direct measurement of grain size and other features is useful. Some core stores are equipped with binocular microscopes; whole core is usually too unwieldy to be examined easily under the microscope, but a close examination may be made of small core chips. Nonetheless, a small torch and hand lens are almost as good as a microscope, and are more manageable. In addition, an acid bottle (for carbonate determination), steel point (to gauge mineral hardness) and grain-size chart are useful accessories (Fig. 4.2).

Proprietary computer software is available for entering data from cores directly onto a laptop (or even handheld) computer. The most basic type simply records a limited range of defined parameters within the core over predetermined depth intervals, and is therefore essentially a digital version of the 'tick-chart' (Fig. 4.3c). Other, more sophisticated, software systems produce a detailed log in which the logger can select from a very wide range of parameters, and in some cases define additional parameters during logging, and also define the depth intervals to be described. The latter type can print out as a multi-columned graphical log that appears little different from the best hand-drawn varieties.

Logged by:
Date:

Sedimentological Log

DEPTH (M)	AGE	FORMATION	IMAGE	LITHOLOGY, GRAIN SIZE AND SEDIMENTARY STRUCTURES Clay, Silt, Sand (VF, F, M, C, VC), Granules, Pebbles, Cobbles, Boulders	REMARKS	SAMPLES	DESCRIPTION	INTER-PRETATION

Section	Sheet of	Interval From: To:	Vertical Scale

Fig. 4.1 *An example of a simple core-logging base.*

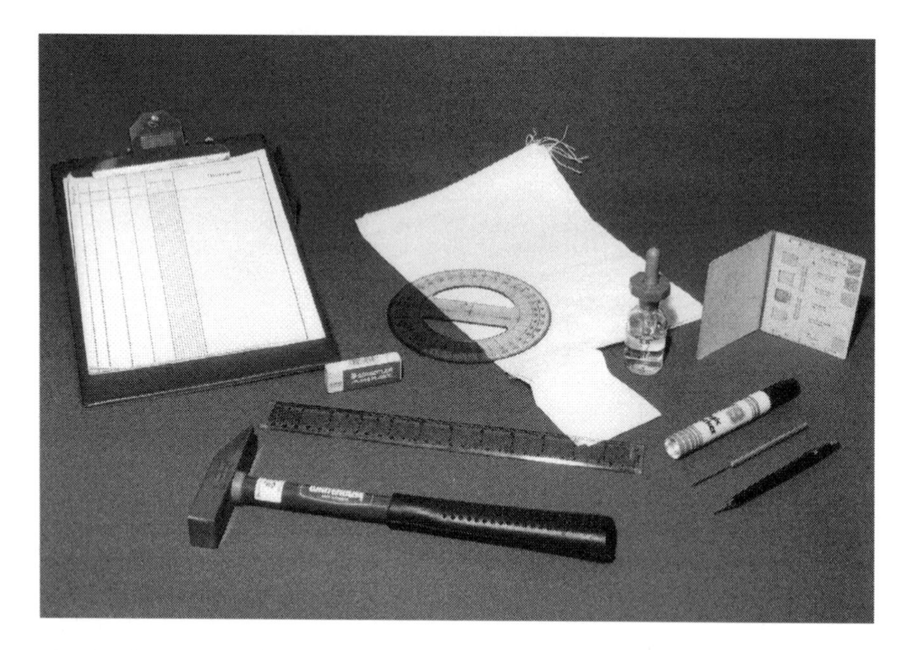

Fig. 4.2 *Core logging does not need a lot of sophisticated equipment.*

These can greatly accelerate logging, and do not of course require any subsequent drafting. The limitation of virtually all of these systems is that, while an indefinite number of features can be defined during logging, the facility for freehand drawing of features is rarely available. Although it could theoretically be incorporated, the accurate visual representation of structures observed in a core on a computer screen requires a data-input method not generally available on a laptop. It could perhaps be approximated using a mouse and mouse mat, but a small digitizing tablet and accompanying pen are required for any degree of precision. The system becomes so unwieldy that it remains simpler to record the information on paper and clipboard. Of course it subsequently can be, and commonly is, digitized.

Another less obvious benefit of a paper log is that the geologist is more inclined to record hunches, alternative interpretations, speculation and miscellaneous comments, which are much less likely to be input to a computer from which a final log for client use will be directly produced. As we have observed above, an informal commentary such as this can be of immense value, especially some time later when the original log is retrieved from a filing cabinet to address a particular problem that may have arisen.

But despite all this, computer input of core data is increasingly being used for relatively rapid and routine logging, where the advantages can significantly outweigh the disadvantages. The precise method applied depends on the nature of both the hardware and the software, in addition to the purposes to which the log will be put. In principle, however, the method of logging does not differ from that used when it is undertaken purely manually, and in view of the potentially wide variety of systems, computer input is not considered further in this chapter.

Many features on a core show up more clearly when wet, so a supply of water may be needed. This is most easily applied with a hand-held polythene spray bottle. If the core is covered with mud, a bucket of water (warm in an ideal world) and a sponge will be needed to wash it down. Care should be taken when making a core wet, however, for several reasons. Some structures are clearer when dry, and water may obliterate them temporarily, or even permanently. When core dries slowly from its natural wet state downhole, a powdery efflorescence of salts sometimes precipitates on the surface. The degree of efflorescence can depend on slight changes in rock properties (such as permeability) from point to point, and can thus highlight details of rock structure that are otherwise indistinct. An over-hasty spray with water can remove this important evidence. Similarly, a rock soaked in oil, either resulting from the use of oil-base mud, or because it represents a hydrocarbon reservoir, should be treated with caution. Addition of water will rarely improve clarity, and may result in the creation of an oil–water emulsion that merely clouds the rock surface for a time. The lesson is always to experiment by wetting a small section of core before applying water more liberally.

Although less common, some cores contain naturally occurring, highly water-soluble minerals such as halite and the potassium salts. Other minerals, such as swelling clays (for example smectites), will be damaged if made wet. Where this is a possibility, water should of course be kept well away until their presence or absence has been firmly established. The visibility of structures in water-sensitive minerals can be enhanced with light lubricating oil if necessary, so long as this will not jeopardize subsequent geochemical or other analyses. The type of oil supplied in aerosol spray cans is ideal for this purpose.

Depending on the precise logging requirements, other equipment will be needed from time to time, but the above is sufficient for general descriptions. The logger will also often need to take samples, in which case a waterproof pen, sample bags and sample cards will also be needed (Sections 3.2.5, 4.3.19).

Once the geologist is suitably equipped, the decision needs to be made as to which end of the core to start logging. There is no strict rule about this. Commonly, logging begins at the top of the first, or highest, core and works downwards. This is chiefly because the drilling sequence is followed, and if logging is undertaken while a sequence of cores is being cut it is the most logical procedure. The top of the core is usually drawn at the top of the logging sheet. If logging were to begin at the base of the core, the precise core length would first need to be checked and measured off on the log in order to achieve this format. Logging from the top, however, disregards the geological convention that rock sequences are described from the bottom in order (generally) to follow the passage of geological time. This is of no particular benefit to the logging geologist, however, and except in special circumstances it is usually more convenient to begin at the top.

Having decided the logging format and recorded relevant details regarding the well and drilling method (Section 4.2), the geologist simply begins to work systematically through the core, section by section, applying geological expertise to create a clear and concise summary of what is seen (not to mention smelt, tasted, felt—and even heard) in the core.

4.2 Format of the core log

4.2.1 Recording and interpreting

Logging core is always a combination of recording details of the core and interpreting what is seen. The scientifically minded will point out, quite correctly, that the geologist should always distinguish between the two, but in practice this can be difficult.

The reason is twofold. First, decisions have to be made from the moment logging begins: whether the depths given by the driller are correct; whether there are missing sections in the core; whether the core has been reassembled in its correct order after falling from the core barrel onto the drilling floor. These decisions depend on the geologist's judgement, and are thus interpretative, yet logging could hardly commence without them. Second, geological terminology is very largely genetic. To some extent this is avoidable. There is no doubt, for example, that the terms 'cross-bedding' or 'inclined bedding' are preferable to 'current bedding'. It is, however, virtually impossible to describe succinctly the characteristic pattern of, say, a *Chondrites* burrow system without simply naming it, and yet this is strictly interpretative. And it is difficult to describe wave ripples in any clear way without implying that the geologist considers they were deposited in an oscillating current.

Although it is difficult to avoid, the geologist should be aware of the extent to which the core is being interpreted in simply making a 'descriptive' log. Of course, many logs will also carry an 'interpretation'. Indeed, there is little point in cutting a core if it is not ultimately to be interpreted. This 'interpretation', which might be of depositional environments in a sedimentary sequence, the economic zones in a mineral deposit, or the inferred planes of weakness in a slope stability study, is best kept quite separate from the basic 'description'. This rule should be followed whatever log format is chosen.

4.2.2 Layout of the log

It must be emphasised from the outset that there can be no generally preferred layout for a core log. There are numerous different types of core, an unlimited number of ways in which a core log can be used, and no end of opinions from geologists (and others) as to how a log should appear. The purpose of this section is not to set rules for logging formats, but rather to explore some of the endless variety of possibilities, with some comments on the different uses to which the varying formats may be put.

Many loggers, especially those working in industry, will have no choice over logging format. Many companies have their own internal standard. Engineering geologists in the UK commonly follow a format based on that suggested by Knill *et al.* (1970) in their Working Party Report on the logging of rock cores for engineering purposes (see Bibliography), and the British Standard *Code of Practice for Site Investigations* (BS 5930:1999) has also now been widely adopted. There is a definite advantage in using a standard format if a suitable one is available, as it is then more easily understood by others familiar with the same standard, and the prime goal of effective communication is thus more easily reached. However, no standard will be suitable for all eventualities over a range of disciplines, and communication is not enhanced by doggedly logging on an inappropriate base.

The format of a log should be looked at from two viewpoints: that of the geologist carrying out the logging, and that of the client, manager or colleague who will be using the final result. Some final logs contain so much information that they are as wide as the desk. This may be useful, but no geologist wants to be inking in a full-width logging form on the deck of a drilling ship in a force 8 gale.

Of course, many of the columns of data in the final log will not be available until completion of laboratory analysis, so the wellsite geologist will not need the full-width logging form for a field description.

Similarly, it is often preferable for the final log of long sections of core to be on a continuous fanfold sheet, especially where the vertical sequence is important to the interpretation. Fanfold logging forms are, however, a nuisance for field use. The sections near to the creases are often difficult to write on, and the paper has a tendency to come apart along the creases, especially if wet. For field use, logging on individual A4 sheets held on a clipboard is usually a simpler solution. The sheets can of course later be spliced together to form a continuous log if so desired.

The choice of scale will constrain the nature of the log, since it largely determines the level of detail of information that can be recorded. There is no theoretical maximum scale, and if considerable detail is needed from a short length of core, there is no reason why it should not be logged larger than life! One of the prime functions of logging, however, is usually to produce a condensed summary, and for most purposes a scale no greater than 1:20 is sufficient to record all pertinent information clearly, and 1:50 is still considered a large scale. At 1:100 it becomes necessary to stylize and merge thin beds and small-scale structures, and some detail is inevitably lost, although the greater conciseness achieved may outweigh the loss. 1:200 is a good scale for summarizing logs of longer cores (10 m or more, say), and is also the standard scale for displaying wireline logs in hydrocarbon wells, allowing a direct comparison of cores with electric log data. At 1:500 only the gross lithological properties can be recorded, and a log at this scale will indicate only the general nature of the core.

For longer lengths of core it is sometimes useful to produce logs at two scales, one detailed (1:20, say), and one to act as a summary (1:200, say). The logging is initially carried out at the larger scale, and the log is later redrawn to the smaller.

Some logging forms allow descriptions and comments to be made wherever the geologist thinks fit, whereas others require descriptions at set intervals. In the former case, the geologist would generally divide the core on the log into sections displaying similar characteristics (the length of individual sections depending largely on the scale of log). Each section would then be separately described. In the latter case, the core would be divided at set intervals (every 0.25 m, say) and each section described. Logging at predetermined intervals is methodical, and ensures that each part of the core is closely scrutinized, but this method does tend to obscure the natural division of the core into lengths of different lithology or structure. This type of logging is sometimes used when it is important to work quickly without missing vital information, or where inexperienced or untrained personnel are carrying out the logging. The concept can be further refined by turning the logging form into a multiple-choice tick sheet, on which at certain depth intervals the logger simply ticks appropriate columns corresponding to lithology, grain size, induration or whatever (Fig. 4.3). This is not only a relatively simple method of logging, but the records can also be readily transferred to a computer database if required. Needless to say, the simplicity and convenience of this method are achieved at the expense of gathering a rather imprecise and incomplete (although not necessarily inaccurate) dataset.

When logging horizontal cores (Section 2.6.4), the originally horizontal structures will in general lie along the length of the core. Because of this, it can be much simpler and clearer to turn the entire log through 90° so that the usual columns become horizontal rows. Assuming that the well-bore was tracking close to the dip of the beds, few bed boundaries will be encountered. If the core was penetrating a cross-bedded sandstone, for example, the entire length of core may display the same structure, although variations in the orientation or precise geometry of the structure will almost certainly occur along its length. This provides a strong argument for drawing the structures of the core accurately, rather than just using symbolic representations. Features such as grain size and levels of cementation can be recorded as usual, although to the geologist interpreting them they will have a different significance from comparable changes in the vertical section.

Four common systems are used to record information on logs. These are graphic symbols, log plots, presence/absence indicators and written descriptions.

One or more of these systems may be used on a given log, and an example using all four is illustrated in Fig. 4.4.

Graphic symbols are commonly used to build up a lithological column, and may be restricted to the basic lithology, or can additionally indicate such features as sedimentary structures, fractures, mineralization and fossils. This composite graphic lithological column is usually the most immediately informative and visually striking section of the log, on which may be 'hung' other data columns and written descriptions. The use of symbols is discussed in Section 4.2.3.

Log plots are essentially graphs of numerical data (which may be estimated or semi-quantitative, as in 'degree of induration' or 'amount of bioturbation'). They can be represented in numerous ways (Section 4.2.4). Perhaps the most common log plot is that of grain size in sedimentary sequences, which is often plotted as one margin of the graphic lithological column, as in Fig. 4.4. Presence/absence indicators are usually contained in a column labelled with a particular feature of the rock for which the presence or absence has particular significance (mineralization, cement, fracturing).

Features that occur at precise depths (such as individual fossils or fractures) can be identified by an asterisk or tick, whereas continuous features (zones of weathering, carbonate cements in sandstone) can be highlighted by shading in the column over the appropriate intervals. It is also possible to vary either the shade or the thickness of the column, or to use differing ornaments to denote varying quantities or pervasiveness of the feature in question (Fig. 4.4). This produces in effect another type of log plot.

Written descriptions are present on most logs. During logging it is wise to make a note of any features encountered, even if their origin or significance is uncertain at the time (see Section 6.6). The initial log must be kept clear and tidy, but there is no harm in its being packed with information. This is not necessarily the case with final drafted logs, on which descriptions should be kept concise, and not crowded; extended written descriptions should generally be consigned to an accompanying report. The aim of a log is to act as an informative but succinct summary of the core, rather than a geological essay. Nonetheless, it is possible with practice to convey detailed information in several short sentences, especially with careful use of abbreviations (Section 4.2.3). It is a useful policy to write descriptions systematically, in a set order, to ensure that no important information is omitted. The range of rock properties recorded in writing and the order in which they are set down will depend on the lithologies cored and the nature of the work being undertaken. For example, the order suggested in the *Sample Examination Manual* published by the American Association of Petroleum Geologists (Swanson, 1981) for sedimentary rocks is as follows:

1 Rock type—underlined and followed by classification
2 Colour

3 Grain size, roundness and sorting, etc.
4 Cement and/or matrix materials
5 Fossils and accessories
6 Sedimentary structures
7 Porosity and oil shows

In contrast, Knill *et al.* (1970) suggest that to engineering geologists, for whom mechanical properties are more important than mineralogy and texture, the rock name is of less significance than other properties of the rock, and should be recorded last. Their preferred scheme is:

1 Weathered state
2 Structure
3 Colour
4 Grain size
 4a Subordinate particle size
 4b Texture
 4c Alteration state
 4d Cementation state

5 Rock material strength
6 Rock name (in capitals)

The British Standard Code of Practice for Site Investigations (BS 5930:1999) differs again, recommending:

(a) Material characteristics:
 1 Strength
 2 Structure
 3 Colour
 4 Texture
 5 Grain size
 6 Rock name (in capitals)

(b) General information:
 1 Additional information and minor constituents
 2 Geological formation

(c) Mass characteristics:
 1 State of weathering
 2 Discontinuities
 3 Fracture state

Note that each of these systems is applicable to the description of a single rock type. Where a description of a section of log covers two or more, perhaps interbedded, rock types, a more general description may be appropriate (depending on the log scale), and it may be unnecessary to give a separate formal description to every lithology.

So the order chosen for any written description will depend on the nature of the rock, and the reason for logging it, but in the absence of a suitable recommended standard, common sense will usually dictate an appropriate sequence. A wider range of rock properties that may be recorded is given in Section 4.3, although this should not be regarded as exhaustive.

A written description will apply to a certain section of the core, which is commonly defined either by a set depth interval predetermined on the logging form, or as a section of core considered by the geologist to form a unit for descriptive purposes. (The exact thickness in this second case will of course depend on the variability of the core, the scale of the log, and the amount of detail required.) When descriptions are required at predetermined intervals, the same description may need to be repeated one or more times. It is generally simplest to write 'as above' (or abbreviated to 'a.a.') so long as the description referred to is immediately above, with no intervening rock units to cause confusion. The geologist should beware, however, of qualifying 'as above', as this can easily cause ambiguities, especially if done several times in succession. For example, a description reading: 'Sandstone with thin shale interbeds, red, coarse grained' may be followed by 'as above but grey', and then 'as above, but lacking shale interbeds'. This is clumsy, and it is unclear in the third description whether the sands are grey or red. It is generally safe to make a minor qualification to a previous description once, but if in doubt it is better to repeat the whole description.

Sometimes it is useful to highlight the similarity (or contrast) between the lithology being logged and that recorded elsewhere (in the same well or otherwise). This is the type of information that may be obvious to the logger at the time, but is much less clear from the log later. A simple note ('Similar to sandstone between 102 m and 105 m') can be invaluable.

There is no need to repeat in written form all the details that are recorded on the various graphic logs. The written description is most effective when it is a concise summary of the most significant features of the corresponding length of core, with an additional note of other details that might otherwise go unrecorded.

The geologist's interpretation of a core is naturally of the utmost importance. However, the final interpretation should not be added until all the relevant information, including laboratory analysis, if appropriate, is available. Notwithstanding this, the geologist should give careful thought to an interpretation while the evidence of the core is at hand. Any interpretation column on the log, or perhaps a separate notebook, should be used at this stage as an *aide-mémoire*; finalization of the interpretation can wait until the geologist is back in the office.

4.2.3 Symbols and abbreviations

The value of symbols and abbreviations has been discussed in Section 4.2.2. They reduce the clutter on a log, and allow more detail to be recorded than would otherwise be possible at any given scale. Like any specialist 'language', however, they can be either an aid or a bar to communication, depending on how and when they are used.

The question that the logger needs to ask before adopting any given set of symbols is whether it will be understood by the final user of the log. Of course, an explanatory legend will occur at the head of all but the simplest logs, or appear in an accompanying report (Section 6.7.4), but if the user needs to refer to the legend for every one of a hundred symbols, the log might as well be written out in longhand.

Fortunately, a few geological symbols, such as the 'brickwork' pattern for limestone and a random pattern of dots for sandstone, are widely recognized. Beyond this, however, there is little uniformity, a fact that reflects not only the perversity of the human race but also the diversity of geological logging requirements. Appendix 1 contains a list of symbols and abbreviations used in this book, which are derived in part from a standard set published by the American Association of Petroleum Geologists (Swanson, 1981). This is useful in that it is moderately comprehensive, at least as far as sedimentary rocks are concerned. An attempt has been made to avoid unnecessary changes from the AAPG standard, although some abbreviations have been converted into British English, and additions and modifications made where, in the author's view, recent changes in geological thinking require them. The list of symbols and abbreviations for metamorphic and igneous rocks has also been supplemented, but is far from exhaustive. Some geological subdisciplines, especially sedimentology, have generated a very extensive list of symbols, of which only a few are included.

Further symbols for use with cores cut for geotechnical and engineering purposes may be found in the British Standard *Code of Practice for Site Investigations* (BS 5930:1999).

Perhaps the best option for any logger to adopt is to try and identify some standard set of symbols defined by his or her institution, industry or company. In the absence of such a standard set, one should be chosen from some other source (such as Appendix 1), and adhered to *so far as is practicable*.

Invention of a personal legend from scratch is not to be recommended, since even if no one but the logger will ever need to read a log, there are endless opportunities for ambiguity, and at least someone else's legend will (one hopes) have been tried and tested in the fire of experience. But having adopted a legend, the geologist should never be afraid *not* to use it in its entirety if, for the intended reader, it is likely to be unclear or ambiguous.

In summary, abbreviations and symbols can be used to reduce text length considerably, and allow rather more detail to be fitted on a log. They are not as widely understood as longhand text, however, and careful consideration needs to be given as to how and where they are used.

4.2.4 Log plots and graphs

Core descriptions comprise a combination of quantitative and qualitative data, which need to be concisely recorded. Much of the quantitative information varies continuously downhole, and is best displayed as some type of graphic plot. The simplest plot is that used to record, for example, grain size, which varies with depth on a horizontal scale, to produce a 'squiggly' line. As noted in Section 4.2.2, the grain-size column is commonly used to form one margin of the graphic lithological column (Fig. 4.4). There are numerous other types of plot. A common requirement is to plot against depth several variables that together add up always to 1 or to 100%. Mineralogical composition is an example, where the sum of the percentage of the various constituents adds to 100%. This is conveniently plotted in a column of constant thickness, divided into 'squiggly' lines running from top to bottom, of which the separation indicates proportions of the various components (Fig. 4.4).

Some types of data are only semi-quantitative. A property such as induration can be plotted on a graph with only (say) three possible values— poorly, moderately and well indurated (Fig. 4.4).

In many cases, data are available only from specific points along the core, rather than continuously. If plotted as simple points on a graph they are unlikely to stand out (Fig. 4.5a). The points may be joined by a line, but this may give the false impression that there are more data points than is the case, and non-existent trends may appear on the log. An alternative format is a histogram in which the values are represented by horizontal bars or narrow rectangles. Sometimes non-continuous data may be best represented by small histograms at the appropriate depths. This might be suitable for displaying the results of grain-size analyses, for example (Fig. 4.6).

Directional data may need to be plotted on a core log. Orientated core, for example, may provide data on the dip and strike of beds. These are commonly plotted as arrows on a graph, on which the direction of the arrow indicates the dip direction (north is up), and the position of the arrow's tail along a horizontal scale indicates the angle of dip, between 0° and 90° (Fig. 4.7). Rose diagrams, containing a summary of directional data, can also be used.

An indefinite number of types of log plot can be devised, but, as always, the one that is chosen should depend on how clearly it portrays the data.

4.2.5 'Field' versus 'laboratory' logging

Some of the practical constraints on logging in the field have already been mentioned in previous sections. Although ideally the geologist might log a core in the format it will eventually have when bound into, or folded into the back pocket of, a report, this is often highly impracticable. In the field, the easiest way of writing is onto A4 sheets, which may be held on a stiff clipboard. These are not only easy to handle, but in wet or windy conditions only a small section of log need be exposed to the elements at any time, so that loss or damage need not jeopardize the whole logging programme. In its final format, however, it may be more convenient for the whole log, or a series of logs, to be drafted for reproduction on a long fanfold sheet or large folding diagram. In general, then, logging sheets for field use should be small and of good-quality, ideally waterproof, paper. It obviously prevents problems later, however, if the final format is kept in mind when designing the field logging sheet, so that unnecessary changes of scale, expansion of column widths and other drafting nightmares may be avoided if possible.

4.3 Core descriptions

Having decided the format of the log, and the range of features to be recorded, the actual description is fairly easy. As always, it is a question of working methodically and carefully. Precisely how the logging is tackled will depend on the nature of the core, the log format, the time available and numerous other factors. Listed below are some of the main classes of information that will need to be recorded. It will not always be necessary to record all of these data items, although some loggers will need to record additional features.

4.3.1 General details

The first information entered on a log will be the name or number of the well, together with its location and any other relevant details. These and related details are usually best recorded on a section designed for this purpose at the top, or on a preliminary page, of the log (Fig. 4.8). The datum from which depths are measured (for example, ground level, 110 m above sea level, kelly bushing) must be recorded. It may be appropriate to record the drilling method, type of core barrel and bit, together with drilling details such as the rate of penetration, loss of circulating fluids or, in shallow wells, standing water levels. In large-scale operations, such as in the oil industry, drilling parameters will usually be recorded in detail by the drilling crew and be permanently available. In smaller operations, especially when using drilling contractors, this will rarely be the case, and the geologist should record all significant information. This can be done effectively only if the geologist remains in close communication with the driller. The differing and sometimes conflicting objectives of driller and geolo-

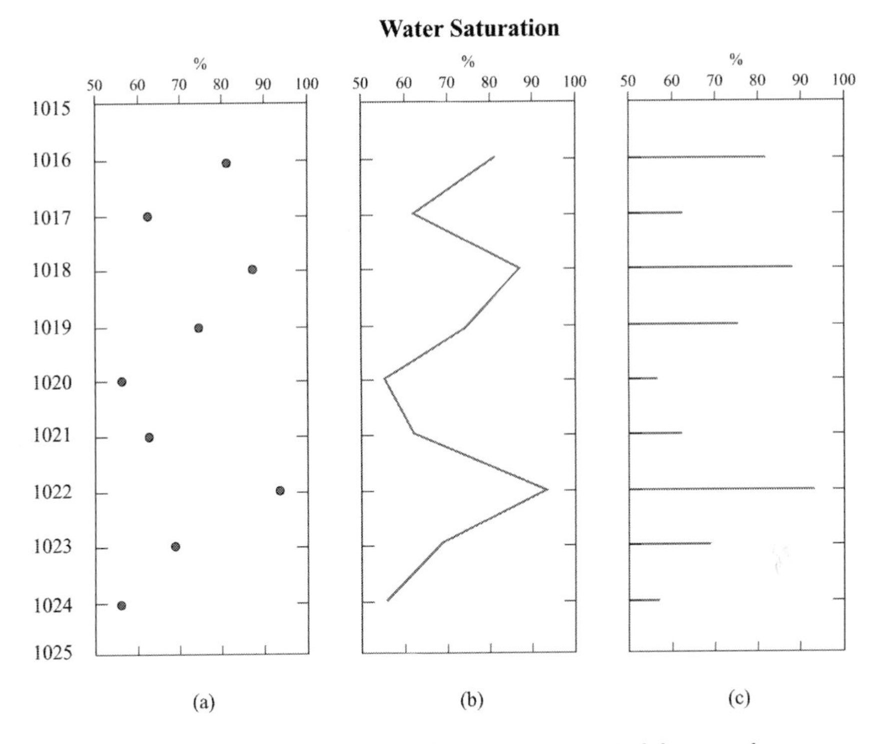

Fig. 4.5 *Three different methods of presenting the same numerical data on a log.*

gist (making holes and gathering data respectively) can sometimes lead to a breakdown in communications. The geologist may need to use tact and diplomacy to ensure that this does not happen.

A note should be kept of the type of core logged (for example, 5-cm-diameter whole-core, frozen slabbed half-core, resinated slabs). Any general comments concerning the state of preservation (such as 'core disordered in places owing to careless handling by inexperienced wellsite crew') are useful. The place and date of logging are required, together with a note of the scale of the log.

A record of difficulties encountered during logging is helpful if the log may have been materially affected by, for example, poor lighting, monsoon, or the lack of proper laying-out facilities. There is a danger, though, that this may emphasize a lack of initiative on the part of the geologist, so it should be used with caution, and should certainly not become a platform for complaining of poor working conditions! Lastly, the author should not forget to add his or her name.

4.3.2 Depths

Before any logging can begin, the depth column on the logging form needs to be completed. The depth of the top of the first core, which may be zero or some

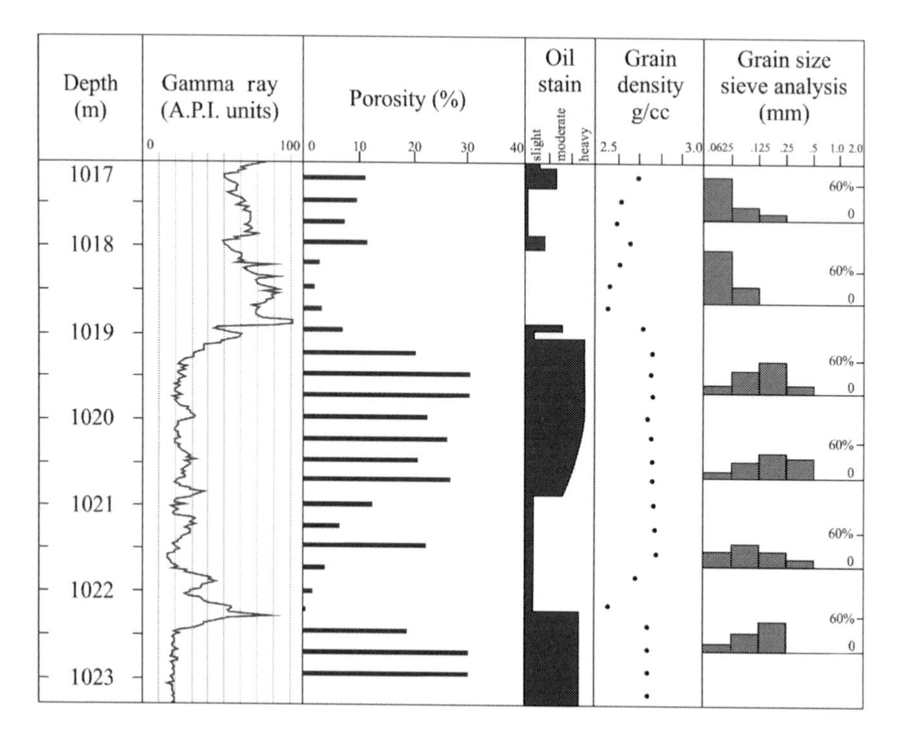

Fig. 4.6 *More ways of displaying data.*

other figure, should have been recorded on the core box (if it has been boxed) in addition to the core itself. This should also be checked against the drilling records, if available. The depth spacing marked on the log will of course depend on the chosen scale. For a 1:20 log the spacing is 5 cm representing 1 m; on a 1:50 scale, 2 cm represents 1 m. Particularly for those geologists using 'imperial' measurements, a preprinted logging form is of considerable help, with 1 foot being represented by 0.6 inches and 0.24 inches respectively at scales of 1:20 and 1:50. As the core is logged, the depths marked on the core itself and the boxes must be constantly checked, since core boxes can easily be laid out in the wrong sequence or upside down. When the base of the core is reached, the depth can again be compared with drilling records, and the length logged should tally with the core recorded at wellsite (Section 3.2.4).

If two lengths of core in a deep well are separated by a large interval, it would usually be most practicable to omit most of that interval on the logging form. It is important that the missing section be clearly marked (Fig. 4.9) so that potential confusion is avoided. If the log is going to be used in a correlation exercise with other wells, or with other data (for example electric wireline logs) from the same well, it is sometimes more convenient to log the cores so that the space between them is to scale, that is, without a missing section. Although this may result in a long 'empty' interval on the log, it does enable one to judge the separation of the

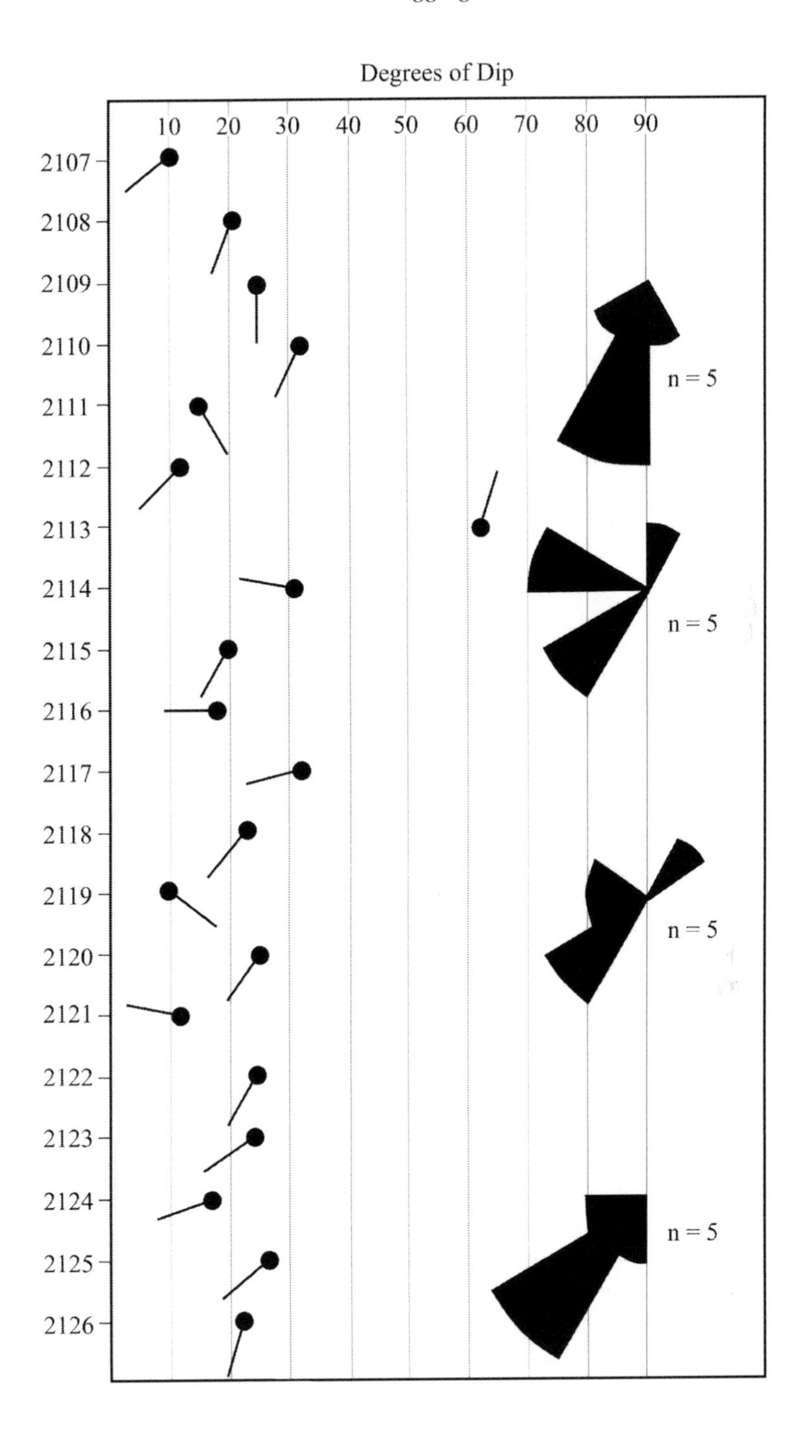

Fig. 4.7 *A plot of varying dip directions in a core. This is sometimes called a 'tadpole' plot. Each tadpole relates to a single reading. Rose diagrams are used to summarize the data.*

CORE LOG

WELL No: 2075-3/2

COUNTRY	U.K.	AREA	Carriden
COMPANY	Edinburgh Exploration Co.	DATUM (a.m.s.l.)	36m
AUTHOR	J. MacTavish	DATE	14th February 2008
SCALE	1:20		
RIG	Drilling 1		
REMARKS	Logged at wellsite on unslabbed wet core		

Fig. 4.8 *Example of a log header, which contains general details of a well, and should appear on the front page or top of a log.*

cores correctly, and it facilitates comparison with other logs at the same scale (Fig. 4.9).

The geologist should ensure that no confusion arises if two depth scales, such as driller's depth and log depth (Section 2.8), are available for a well. In deep wells, especially in the hydrocarbon industry, many wells deviate considerably from the vertical, and the depth measured downhole can differ substantially from the true vertical depth (TVD). This presents an additional potential depth scale, in which 1 m of depth corresponds to more than 1 m of core. To add to the complexity, TVD is usually measured from sea level, even if driller's and log depths are measured from some arbitrary datum such as the kelly bushing (Section 2.2). Core is conventionally measured according to driller's depth, and if there is any possibility of confusion the depth column must be clearly marked as referring to driller's depth. It may sometimes be useful to include also a scale of log depth or TVD on a log, in which case these also must be plainly marked (Fig. 4.10).

In the petroleum industry reference is often made to 'measured depth'. Unfortunately, some operators use this term to refer to the driller's depth, and others to refer to the log depth (with the balance swinging perhaps from the former to the latter over recent decades). This type of confusion of terminology makes life interesting for the geologist, but so long as he or she is awake, it rarely causes a real problem.

Depths on logs of horizontal cores (Section 2.6.4) will be marked initially according to the driller's depth, in the standard fashion, with the usual possibility of a shift from driller's depths to log depths. It becomes interesting when true vertical depths are marked. In principle a 100-ft core may all lie at a single value of TVD, and the depth marker may run along the length of the core. In practice, even so-called horizontal cores will meander up or down a little in the vertical section. It is unlikely that the geologist will be able to work out the TVDs for him or herself, and will have to rely on the well survey results.

4.3.3 Core position

The top and base of each core, together with the number of the core, are valuable information on the log (Fig. 4.10). They may be shown by ruling a heavy horizontal line across the log, and a note ('Top of core number 3') in the description column. This is particularly important where the base of one cored section coincides with the top of the next, so that there would otherwise be no break in the log. The tops and bases of cores are particularly vulnerable to damage, and spurious lithologies falling downhole between coring runs can

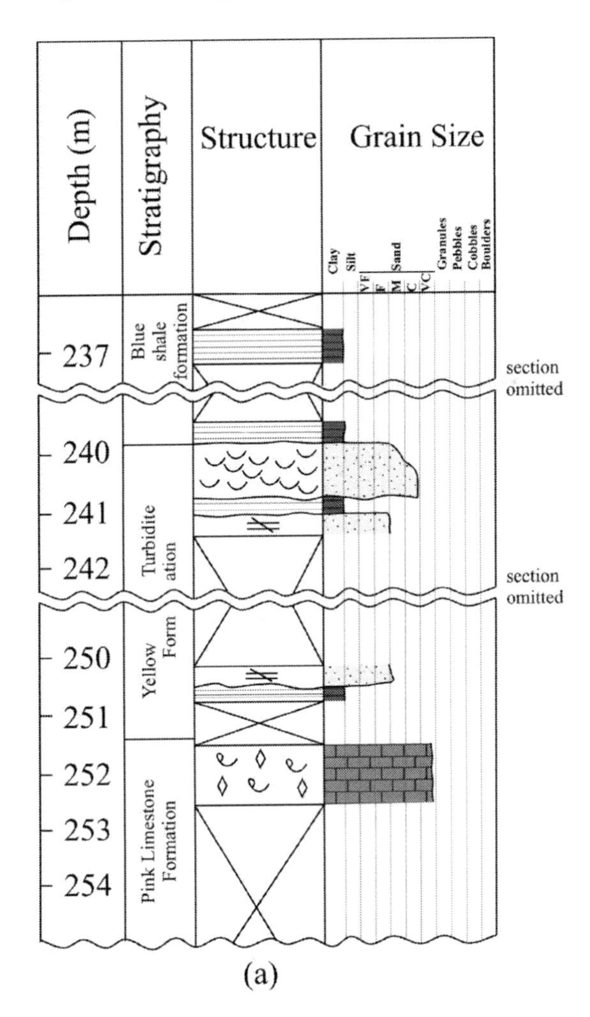

(a)

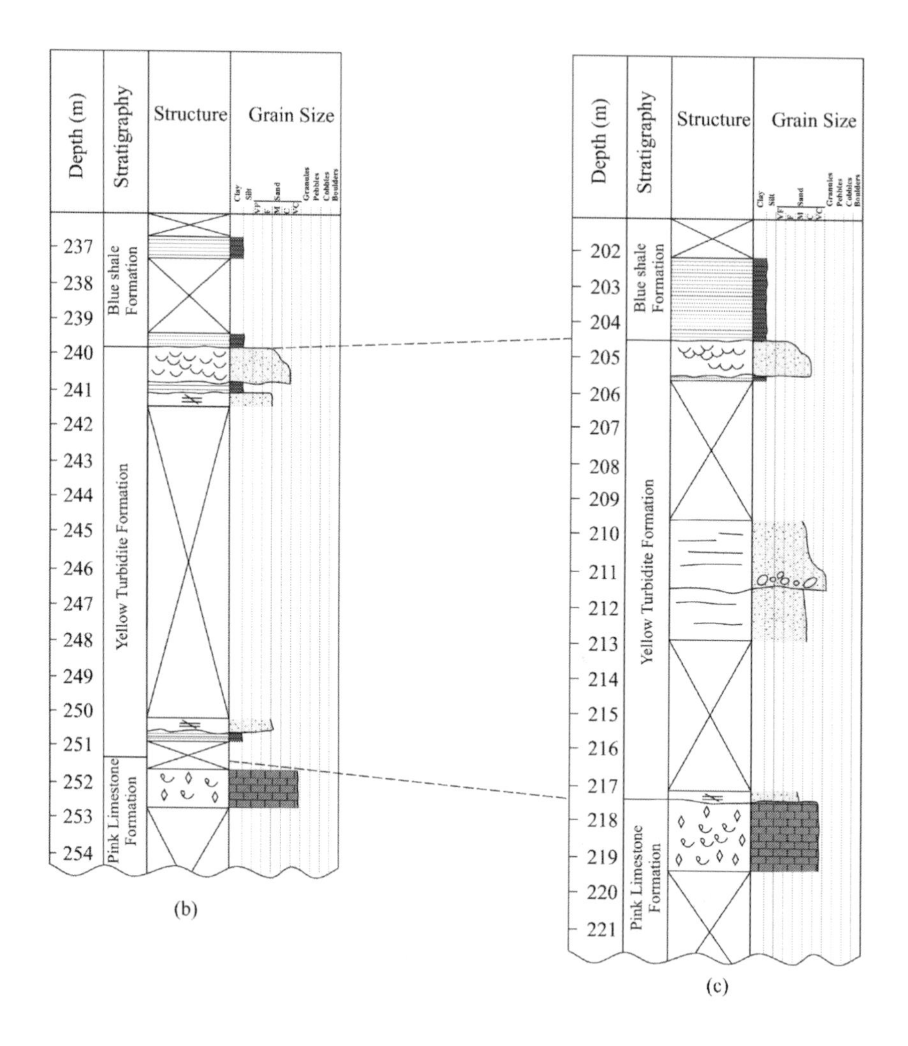

Fig. 4.9 *Missing sections of core (either where a core was not cut, or where it was lost) must be marked clearly. In this case, a pair of crossed lines is used. Where long sections of core are missing, a 'non-scale gap' may be left in the log (a), so long as this is clearly marked. In some cases, as in (b) and (c), where sections of two or more wells are being correlated, it is preferable to leave the missing sections at their proper scale so that the relationships between formation thicknesses are apparent.*

become incorporated in the cored section and logged. This information can therefore be valuable in ensuring a correct interpretation.

4.3.4 Rock name

Ideally the rock name would belong to some formal system of classification. Unfortunately, these mostly require thin-section petrography, since few geologists would recognize a kentallenite or an intrabiopelmicrite at wellsite. It is best to give a rock name in which one is confident, so 'basic igneous rock' or 'limestone' is usually quite sufficient. If necessary, this can be amended following subsequent laboratory analyses. If one is still uncertain whether, for example, a rock is a limestone or a calcareous sandstone, then it can simply be logged as 'limestone or calcareous sandstone'. This is much safer and more honest than using guesswork, and no one should criticize a genuine expression of uncertainty.

4.3.5 Colour

For general purposes one can use everyday colour descriptions: reddish-brown, buff, greenish-grey, etc. These are not very specific, however, and each of these terms can actually describe a multitude of different shades. If more precision is desirable, a rock colour chart such as that published by the Geological Society of America (Goddard *et al.*, 1948) can be used. This comprises a collection of colour 'chips', like those appearing in paint catalogues, in which each colour is defined by three attributes known as 'hue', 'value' and 'chroma'. A reddish-brown colour might thus be designated '10R 3/4', which means that the 'hue' is 10R, the 'value' is 3 and the 'chroma' is 4. Each colour chip is also designated by a (non-unique) name such as 5YR 6/1: 'light brownish grey'. If one is using these names, it is important to make clear that they are derived from the GSA chart, since they include some oddities. The 5YR 6/1 'light brownish grey', for example, to many people appears distinctly pink!

The colours of the GSA rock colour chart are based on the Munsell notation. A similar but much more comprehensive set of colour charts is sold by the Munsell Color Company of Baltimore, USA. Various charts are available from this source, covering every colour under the sun. The most useful to the geologist, however, are the Munsell Soil Color Charts, which are sold in a compact loose-leaf binder. As their name implies, they are designed for defining soil colour. Although they also cover most rock colours, it is in fact chiefly in describing cores of soils that precision in recording colour is important, since many soils are classified by colour. With rock successions, it is generally unnecessary to go to such lengths.

Some rocks comprise two or more colours, either because they are composed of different minerals or clasts, or because there is some type of mottling due to weathering or diagenesis. In this event, the separate colours should be recorded, together with a description of what is controlling the colour variability. If it is green mica in a generally white granite the reason is obvious, but sometimes colour banding or mottling may be present for no clear reason. It may, for example, comprise patches of reddening in an otherwise green sandstone. A sketch of the colour distribution can be made in the description column, or a graphic lithology column suitably

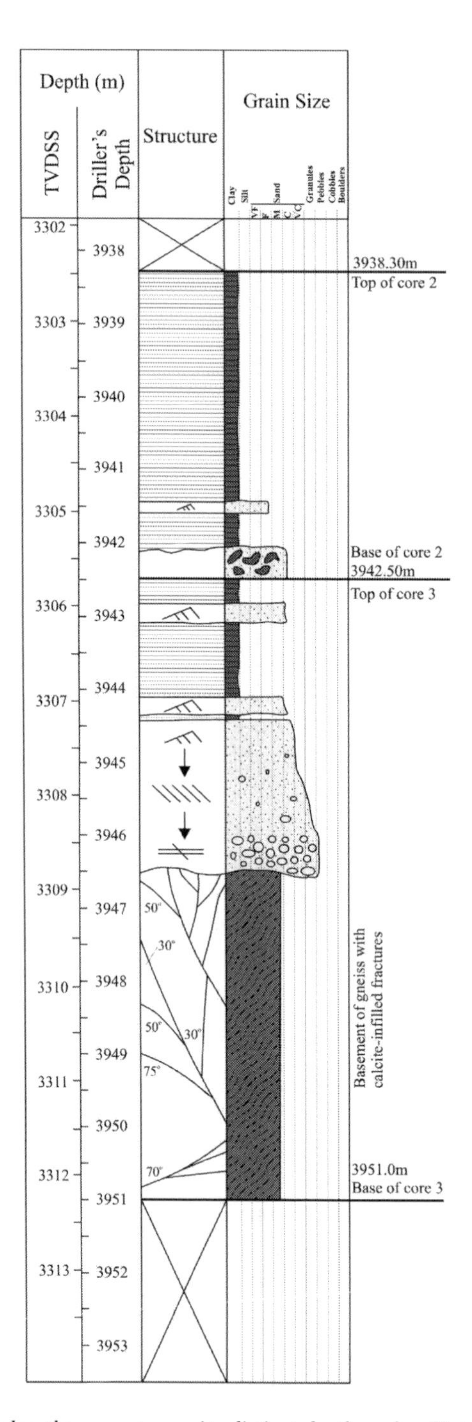

Fig. 4.10 *On this log there are two quite distinct depth scales. Note also the marking of tops and bottoms of individual cores, and the annotation of basement fractures with their true angular relationships, to assist structural or geotechnical interpretation.*

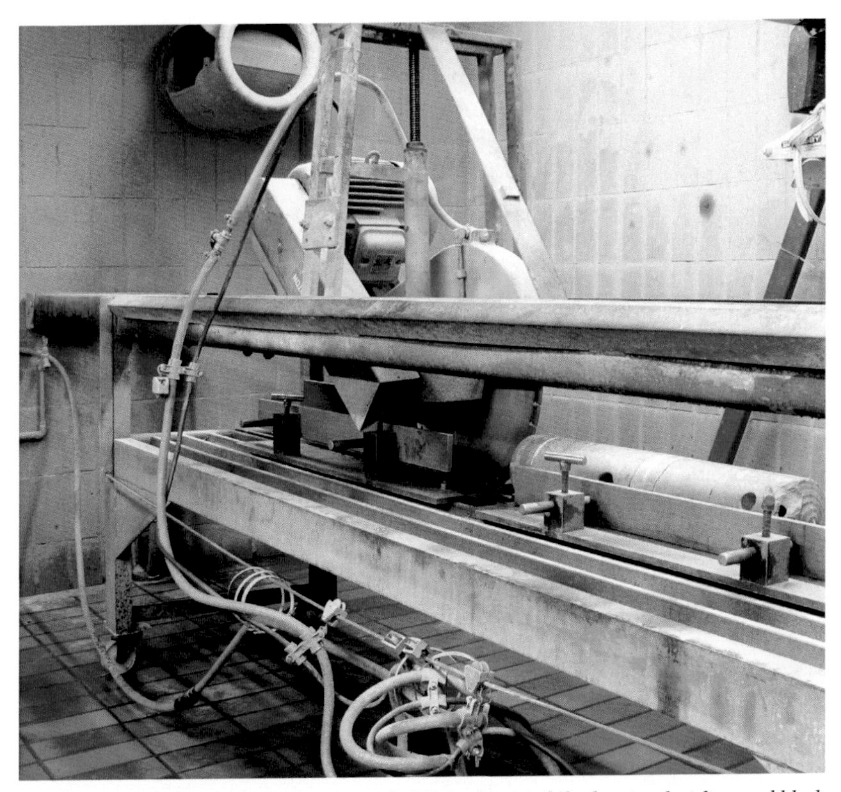

Fig. 3.7 *Core being slabbed. The core is held in a frame while the circular diamond blade moves along its length. (Photo courtesy of Core Laboratories Ltd.)*

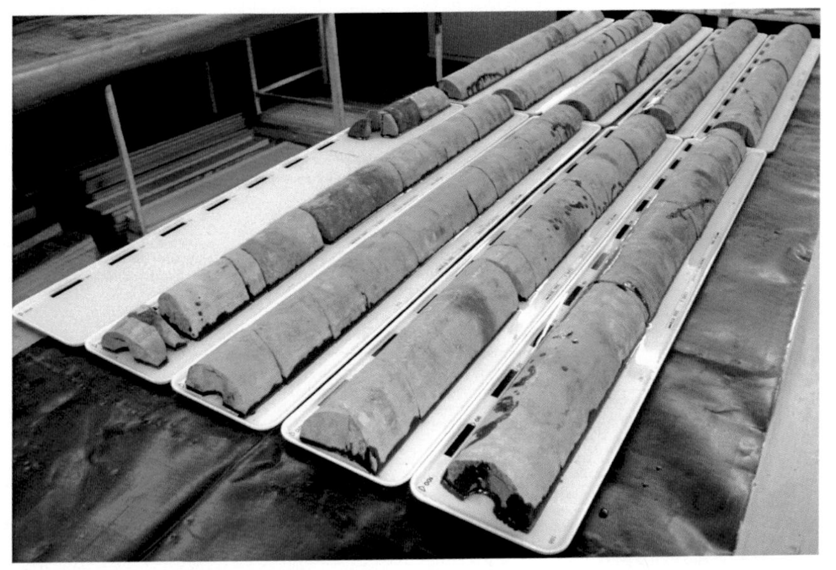

Fig. 3.9 *Half-core lying in trays, now embedded in resin. Once the resin has set, the core remaining above its surface will be cut off, to leave smooth, parallel-sided slabs. (Photo courtesy of Core Laboratories Ltd.)*

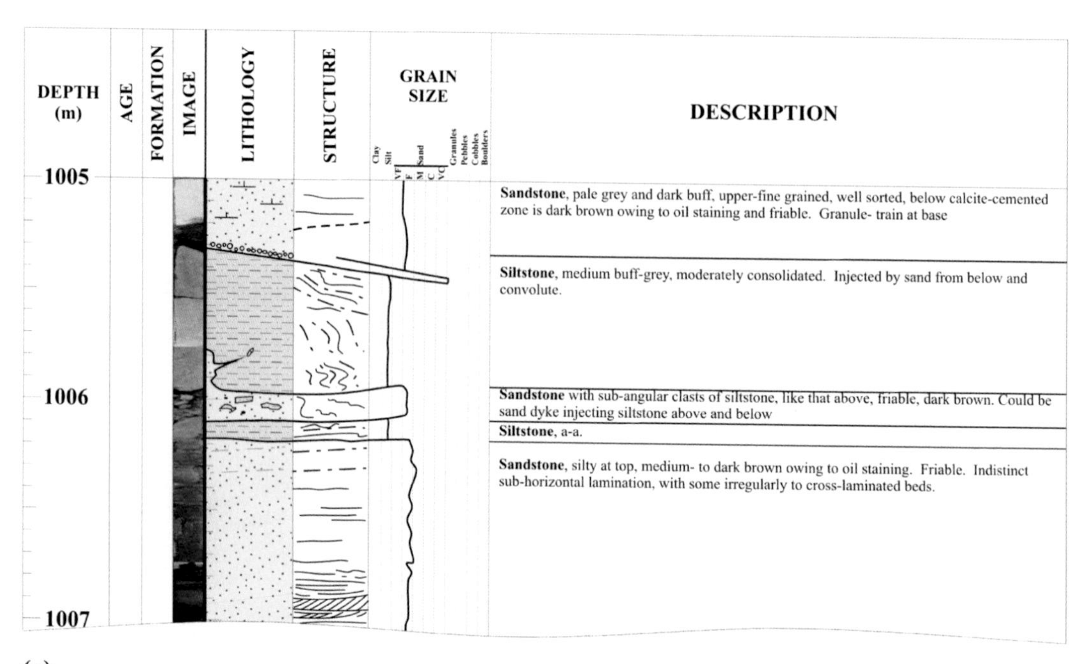
DEPTH (m)
AGE
FORMATION
IMAGE
LITHOLOGY
STRUCTURE
GRAIN SIZE
DESCRIPTION
1005
1006
1007
Sandstone, pale grey and dark buff, upper-fine grained, well sorted, below calcite-cemented zone is dark brown owing to oil staining and friable. Granule- train at base
Siltstone, medium buff-grey, moderately consolidated. Injected by sand from below and convolute.
Sandstone with sub-angular clasts of siltstone, like that above, friable, dark brown. Could be sand dyke injecting siltstone above and below
Siltstone, a-a.
Sandstone, silty at top, medium- to dark brown owing to oil staining. Friable. Indistinct sub-horizontal lamination, with some irregularly to cross-laminated beds.

(a)

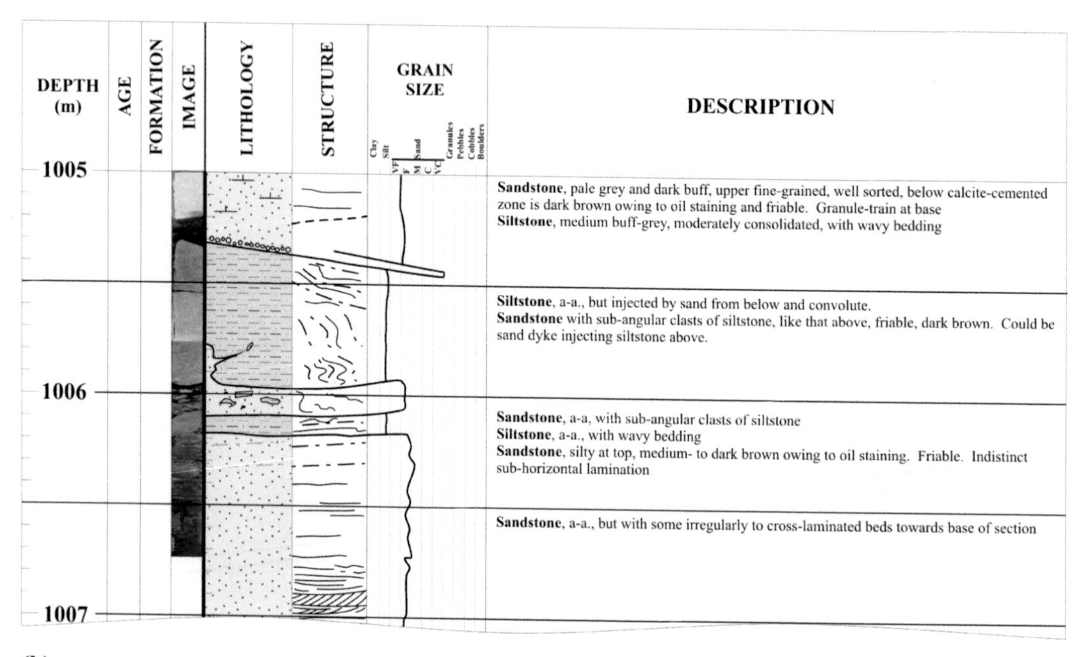
DEPTH (m)
AGE
FORMATION
IMAGE
LITHOLOGY
STRUCTURE
GRAIN SIZE
DESCRIPTION
1005
1006
1007
Sandstone, pale grey and dark buff, upper fine-grained, well sorted, below calcite-cemented zone is dark brown owing to oil staining and friable. Granule-train at base
Siltstone, medium buff-grey, moderately consolidated, with wavy bedding
Siltstone, a-a., but injected by sand from below and convolute.
Sandstone with sub-angular clasts of siltstone, like that above, friable, dark brown. Could be sand dyke injecting siltstone above.
Sandstone, a-a, with sub-angular clasts of siltstone
Siltstone, a-a., with wavy bedding
Sandstone, silty at top, medium- to dark brown owing to oil staining. Friable. Indistinct sub-horizontal lamination
Sandstone, a-a., but with some irregularly to cross-laminated beds towards base of section

(b)

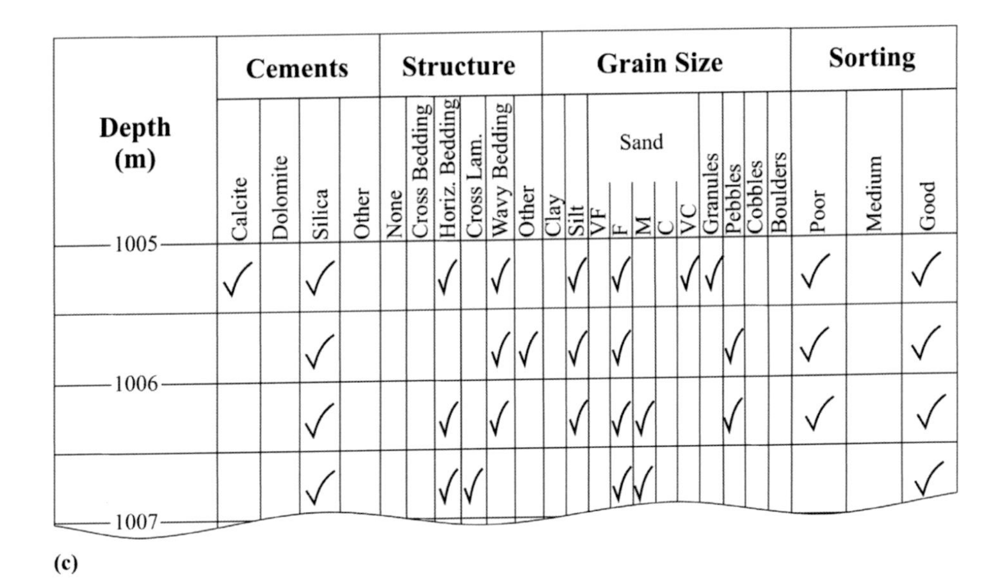

(c)

Fig. 4.3 *Three different methods of logging the same section of core. (a) This produces the most 'natural'-looking log. The logger mentally divides the core into units of similar lithology or facies (or which contain a uniform gradation between lithologies), and describes each such division separately (either in symbols or in words). To log in this way will generally require sufficient time and space to lay out the core and decide on the appropriate lithological division before logging begins. (b) This is more suitable for rapid logging (e.g. at the wellsite), or for logging by less experienced personnel. Preset lengths of core (0.5 m, say) are inspected and described in turn. Each length is given its own description, without necessarily any reference to what is above or below. In theory, the log should be very similar to that produced in (a), but in practice the technique can obscure the natural lithological breaks and transitions in the core, especially where there is a gradual transition over a length greater than that of the individual lengths being described. In both (a) and (b), a scaled-down photographic image of the core has been spliced into the digital version of the log. (c) This comprises a tick-chart, which divides the core into preset lengths like (b), and records a range of parameters in the core. The results are very imprecise, but are useful when very rapid logging of long lengths of core is required. The data may readily be fed into a computer database, e.g. for statistical analysis.*

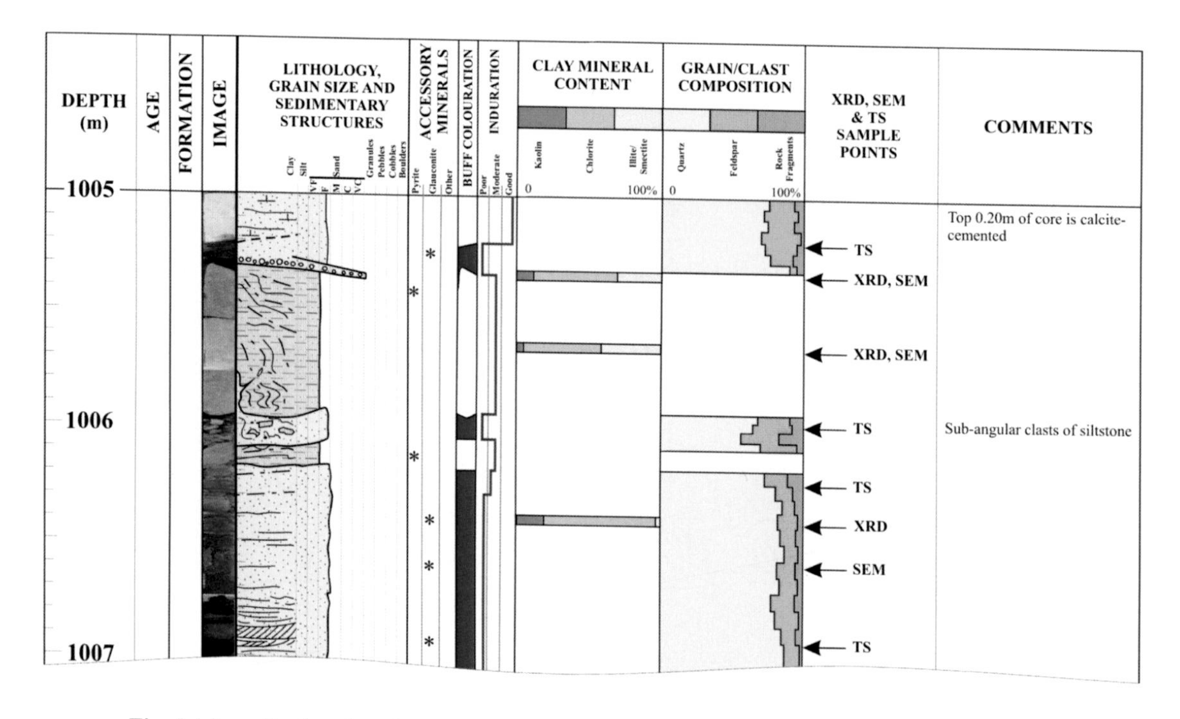

Fig. 4.4 *Log of a short length of core, showing various different ways of presenting data.*

Opposite: **Fig. 4.13** *Core log of the type recommended by Knill et al. (1970) for engineering purposes. The use of colour is optional.*

DRILLING METHOD: Rotary auger to 3.6m Rotary coring with water flush to bottom of hole (12.7m)	GROUND LEVEL: +103.7M O.D.	CO-ORDINATES OR GRID REF: NT 2297 7347	BOREHOLE No: 3
MACHINE: Ga35 / CORE BARREL DESIGN AND BIT: BWF Diamond bit	ORIENTATION: Vertical	SITE: Home Farm	

WATER PRESSURE TEST cm/sec x 10⁻ (1, 10, 100) | WATER RETURN % & LEVEL (20, 60) | DRILLING PROGRESS | CASING | DISCONTINUITIES | FRACTURES per m. (4, 16) | CORE RUNS DIAMETER & DEPTH (m) | CORE RECOVERY % (20, 60)

Drilling progress: 20.11 07; 21.11 07; 22.11 07. Water levels: 21, 22, 23.

Core runs: 42mm; 42mm; 42mm. Depths: 1, 2, 3, 4, 5, 6, 7, 8, 9, 10, 11, 12, 13.

Discontinuities	Attitude
Clay-filled fractures in weathered zone	⊥
Bedding-plane fractures, variably open or clay-filled	⊥, ∠
Uniformly-spaced open planar joints	⋎, ∠
Bedding plane fractures. Haematite- and limonite stained	∠
0.55m thick shattered zone	V, ⋎
Randomly-distributed bedding-plane fractures	∠

Sample	DESCRIPTION OF STRATA	O.D. LEVEL	LOG
0 1, 0 2, 0 3	BOULDER CLAY: sandy clay with pebbles and cobbles of sandstone and a few Carboniferous limestone erratics 3.60	100.10	
	SANDSTONE: grey, fine-grained, micaceous, with siltstone bands and laminae 5.12	98.58	Mc
	SILTSTONE: brownish-grey, micaceous, sandy in part; gradational base 6.76	96.94	Mc
	SANDSTONE: light-grey, fine to medium-grained, micaceous, with silty partings bioturbated 9.10	94.60	Mc
	MUDSTONE: grey, silty, micaceous, with ironstone bands and nodules 10.45	93.25	
	SANDSTONE: light brown, thick-bedded, coarse-grained, feldspathic 12.70	91.0	
	Bottom of hole		

EXPLANATION:

- □ U100 sample
- 0 Disturbed sample
- ■ Core sample
- W Water sample
- 22 Day
- ▽ Ground-water depth first encountered
- ▼ Morning water level
- 21.11.07 Depth of borehole
- ⊥ 80°-90° — Attitude of prominent fractures
- ⋎ 60°-80°
- ⋎ 30°-60°
- ∠ 0°-30°
- — — Solid core recovery
- —— Total core recovery

REMARKS:

LOGGED BY: J. MacTavish	SCALE: 1:50
CONTRACTOR: Carriden Drilling	CLIENT: Edinburgh Exploration Co.	REF. No.:

Fig. 5.2 *A core plugging drill being used to cut a series of 25-mm plugs. (Photo courtesy of Core Laboratories Ltd.)*

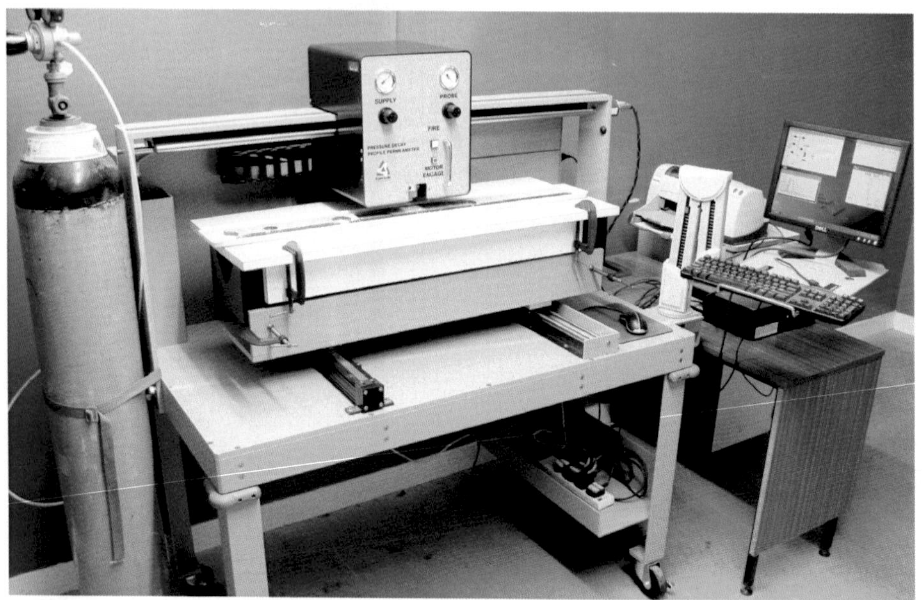

Fig. 5.4 *A probe, or profile, permeameter designed to measure permeability at any position on a core slab. It measures the flow of gas injected at pressure into the rock from a fine-tipped probe pressed against it, and is able to build up a two-dimensional profile of the permeability variations across the core surface. (Photo courtesy of Core Laboratories Ltd.)*

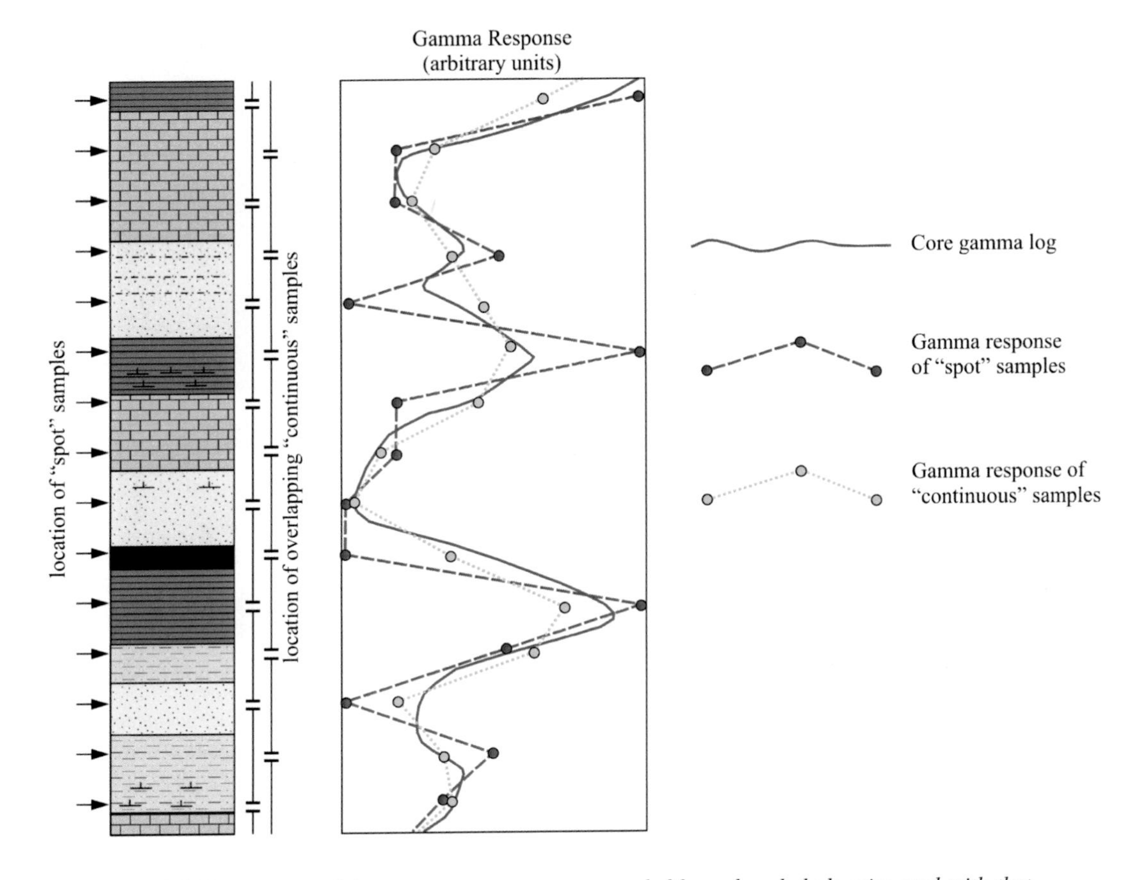

Fig. 5.5 *Comparison of the gamma-ray response recorded by a downhole logging tool with that from core samples over the same section. Because of a varied lithological succession (left), the gamma response varies widely. The downhole tool measures the response over an interval, and because of the relatively thin bedding, the response recorded is never that of a single lithology, but appears smoothed out. Measurement of the gamma response of spot samples of core, however, includes extremes of high and low emission, and produces a 'spiky' plot showing a poor correlation with the downhole response. By taking 'continuous' core samples (see text), the correlation may be improved.*

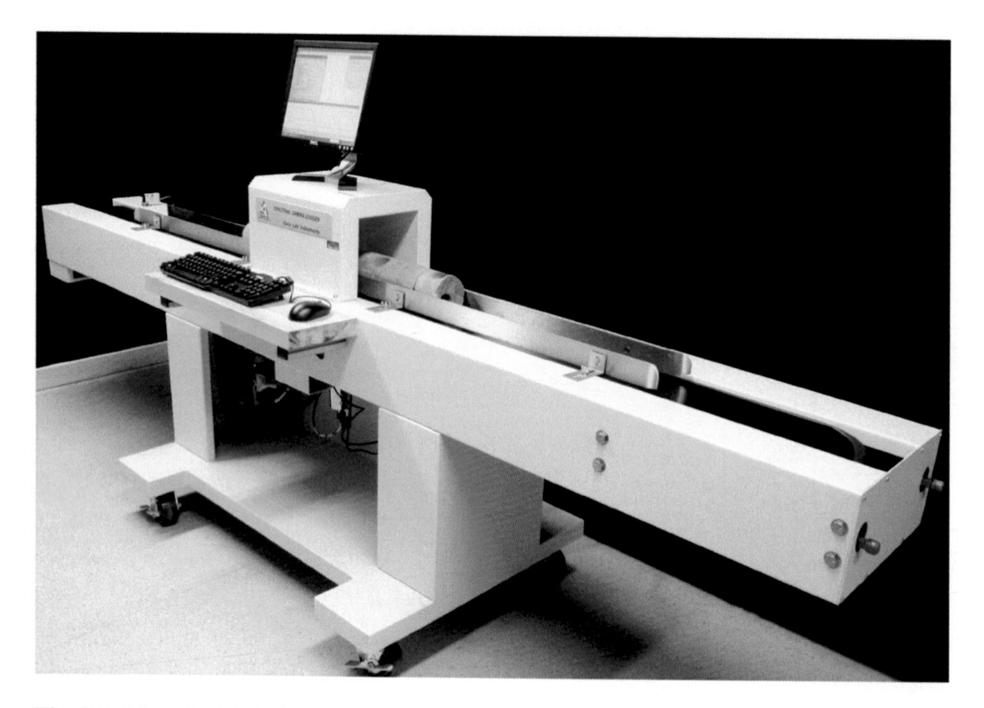

Fig. 5.7 *A length of full-diameter core passing on a conveyor past a core gamma spectrometer (housed within the 'bridge' over the core). The results appear on the monitor screen. (Photo courtesy of Core Laboratories Ltd.)*

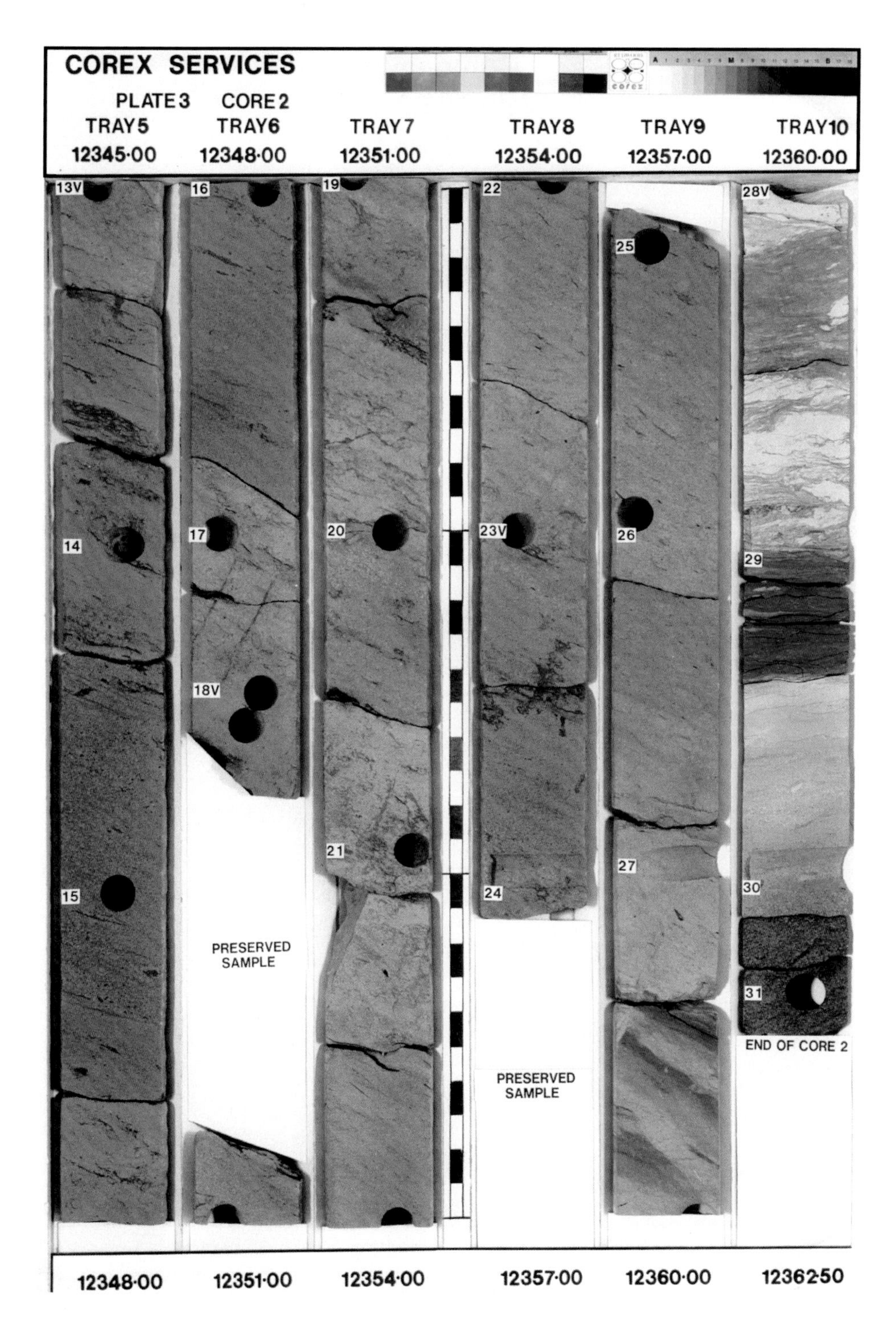

Fig. 5.8 *A good-quality core photograph taken from resinated slabs, which may form an excellent long-term record of a core, complementary to the log. (Photo courtesy of Corex Services Ltd.)*

Company

Well: 00/00-00

Core 1 Plate 1 of 100

1000.00 - 1001.00 m

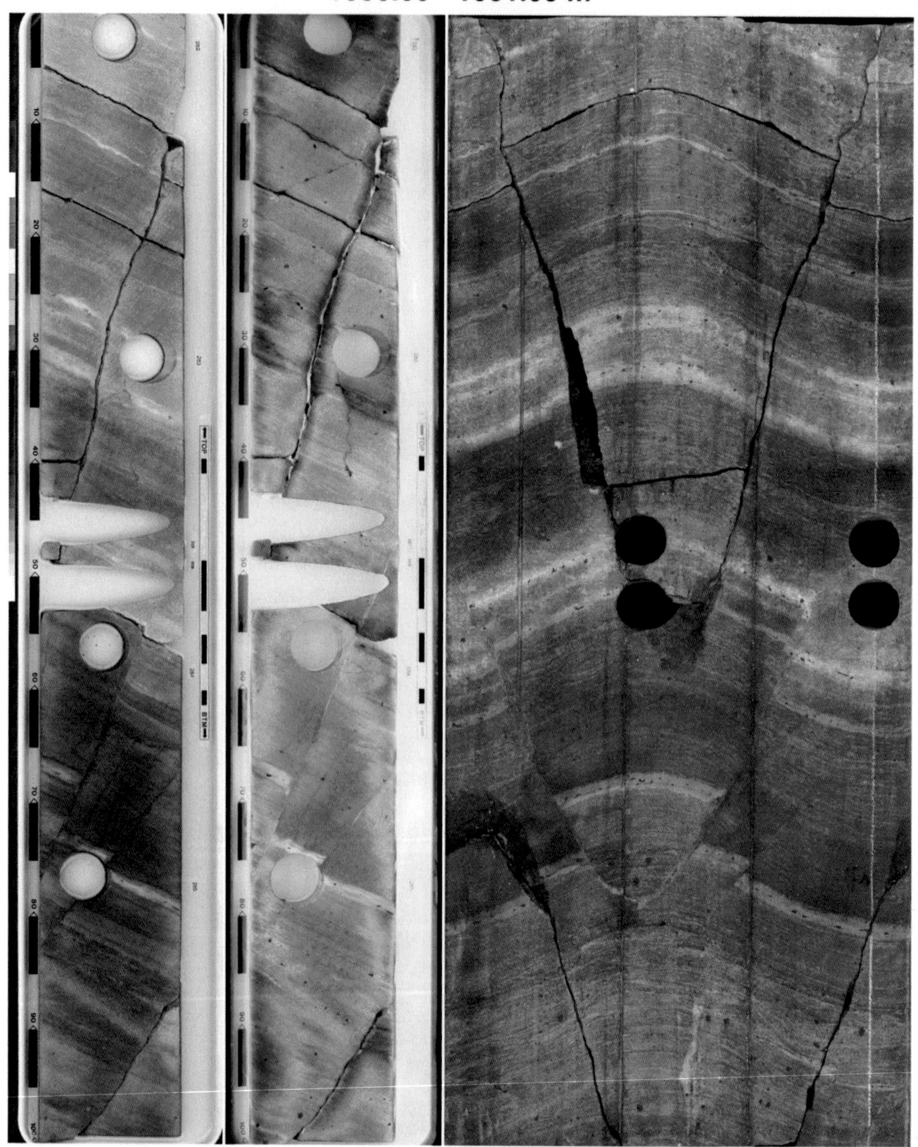

Fig. 5.9 *Photograph of the same 1-m length of core taken in normal (left) and UV (centre) light; on the right is an image representing the entire 360° surface of the equivalent section of whole core. (Photo courtesy of Core Laboratories Ltd).*

Fig. 5.12 *Cleaning samples by Soxhlet extraction. Organic solvent, which is boiled in the flask at the bottom of the apparatus, condenses in the chamber at the top. This drips down onto the plug samples in the central chamber, removing hydrocarbons from them, before returning to the lower flask. (Photo courtesy of Core Laboratories Ltd.)*

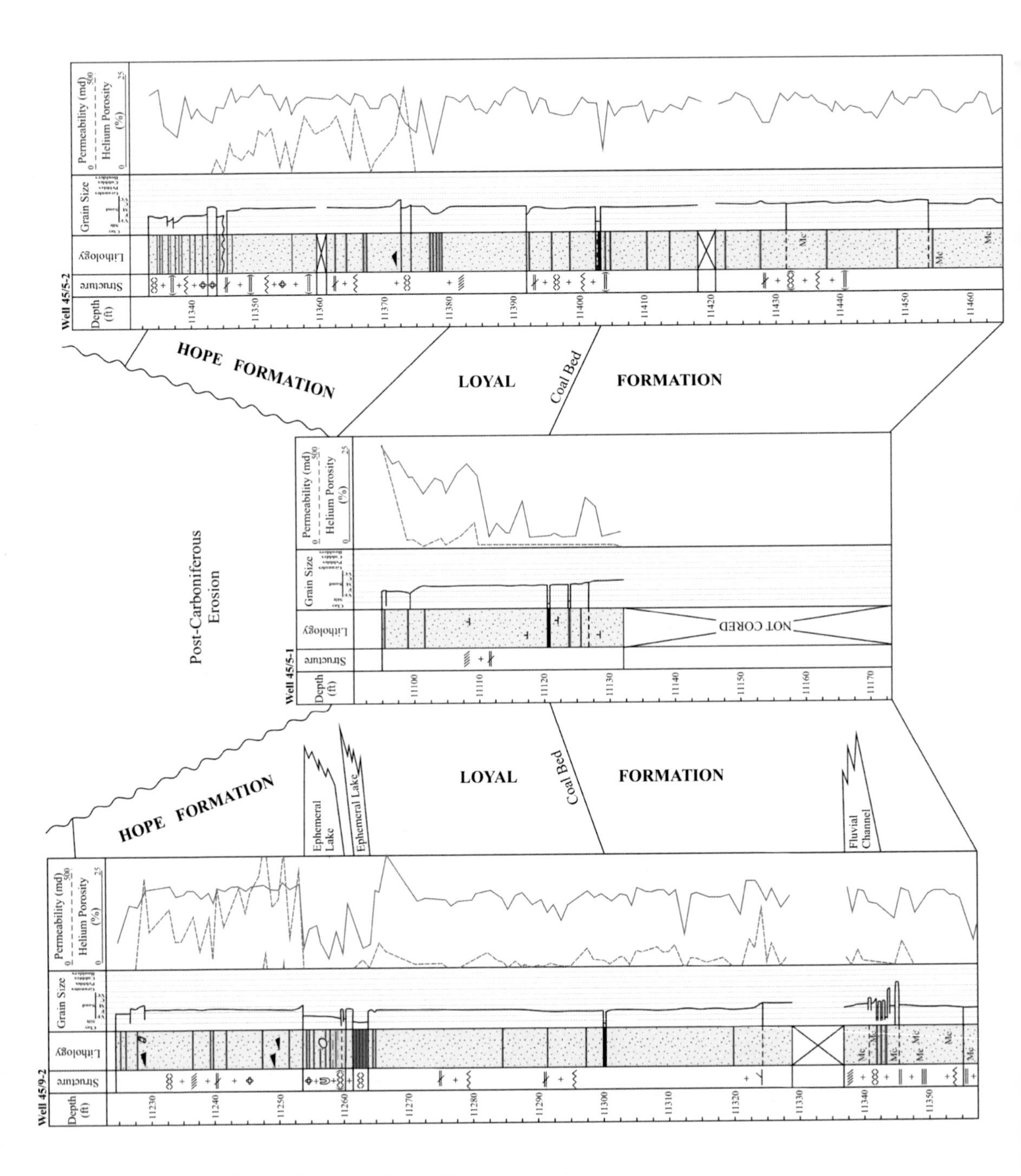

Fig. 6.2 *Core logs used in a correlation diagram.*

annotated, since such subtle clues are sometimes invaluable in geological interpretation.

4.3.6 Fabric

Fabric includes factors such as grain size and, in clastic sediments, roundness, sorting and sphericity. With a carbonate rock the fabric may be, say, oolitic or vuggy, whereas in an igneous rock this would include the texture, such as porphyritic or spheritic, and with metamorphics it may be schistose or banded. In fine-grained lithologies some of these fabrics will not be discernible during logging, and their recognition must await laboratory analysis.

Grain parameters such as roundness, sorting and sphericity are best judged using charts such as those published in sedimentary petrography handbooks: see for example Swanson (1981) in the Bibliography. This ensures some degree of uniformity between geologists. Grain size in clastic sediments is similarly judged by a chart, which may either comprise a suite of sieved sand- and silt-sized fractions mounted on card or in plastic, or may be a representation of different size fractions printed on plastic. One can with practice learn to judge these parameters unaided, but it is still worthwhile to confirm one's accuracy regularly on the appropriate chart.

4.3.7 Appearance, texture and strength

This refers to the physical characteristics of the core in hand specimen, which may include plasticity, fissility and degree of induration. Indeed any distinctive properties of the core that are not noted elsewhere may come under this category. Does it make a distinctive sound when hit with a hammer? Does it break with a conchoidal fracture? Does it smell? Geologists over the decades have used any number of rock properties to characterize or correlate different lithologies. There is, for example, a type of coal ('Parrot coal') that squeaks when burnt! One should not be afraid to record even unusual features—they may turn out to be important, or at least useful, and if not, nothing has been lost.

While the degree of induration for some purposes is of passing interest, or is more important as an indicator of some other parameter (such as cement proportion), for engineering geologists it may be the most important factor. Although, in this situation, laboratory measurements will be needed, British Standard BS 5930:1999 does describe an *ad hoc* field test related to uniaxial compressive strength, based on the following:

Very weak:	Gravel-size lumps can be crushed between finger and thumb.
Weak:	Gravel-size lumps can be broken in half by heavy hand pressure.

Moderately weak:	Only thin slabs, corners or edges can be broken off with heavy hand pressure.
Moderately strong:	When held in the hand, rock can be broken by hammer blows.
Strong:	When resting on a solid surface, rock can be broken by hammer blows.
Very strong:	Rock chipped by heavy hammer blows.
Extremely strong:	Rock rings on hammer blows. Only broken by sledgehammer.

The point is made, of course, that the indications given will depend on a variety of variables, not the least of which would be the strength of the logger, who may need to calibrate his or her own strength against rocks of known strength.

4.3.8 Cement and matrix

Cement is any material that binds a sediment together, whereas matrix is relatively fine-grained material lying between coarser grains or particles. Cement and matrix are by definition not the major component of a rock; a recrystallized limestone is not considered as 100% cement, and a mudstone is not a matrix unless it contains a substantial proportion of larger clasts. Nonetheless, these can contribute significantly to rock properties, and must be carefully recorded. If the core being examined was cut in a well from which wireline logs are also available, zones of cement will often be apparent owing to the decrease in porosity and increase in density that they cause, which may be apparent on, for example, density/neutron logs or sonic logs.

Cement is generally some authigenic mineral, which may or may not be the same as that of the grain being cemented. For example, a quartz sandstone may be cemented by silica, which either comes from almost *in situ* solution and reprecipitation, or may be derived from some more distant source. A calcite cement in the same quartz sandstone would probably have been transported from elsewhere, although it may have been sourced from original shelly debris within the sand, which was subsequently completely dissolved.

Almost any mineral may cement a sediment, and other substances such as crude oil or bitumen may also bind a rock together. The simplest cement to identify is calcite, which will react strongly with dilute hydrochloric acid (remember to wash surplus acid from the surface of the core with plenty of water to prevent the core from being permanently stained and defaced). Quartz cements will generally be evident from the characteristic euhedral overgrowth visible through the hand lens on a broken surface, although this may be less clear on a more indurated sandstone where lack of porosity may have prevented quartz euhedra from forming. Among the other minerals that may cement a rock, some cannot be readily identified in hand specimen, and must just be described, with subsequent laboratory analysis if necessary.

Graphic symbols exist for all common cements (Appendix 1), and can be used on the graphic lithological log. The cement type should also be included in the written description, with a note as to the nature of its occurrence. It may, for example, be pervasive, patchy, poikilotopic, replacive or nodular. In some cases the cement may have a genetic or environmental significance, as with 'cornstones' (fossil calcrete) or septarian nodules. If an interpretation of depositional environment is important, these features should be carefully described.

Matrix is simply a sediment type, and is described in the same way as any other sediment. Its relationship to the larger grains or clasts it contains may need to be stated. For example, does the matrix support the individual grains, or do the grains support each other by mutual contact, with the matrix just forming a passive infill? If the latter case, is there evidence as to whether the matrix was deposited at the same time as the larger particles, or was it infiltrated between them at a later stage? With coarse conglomerates, it is possible to draw accurately the relationship between the grains and matrix in the graphic lithological column.

4.3.9 Fossils

This includes not only the type and number of fossils, but also their state of preservation and disposition within the sediment. Depending on the aim of coring, fossils can be among the most useful indicators encountered while logging, giving information on depositional environment, age and sediment transport mechanisms.

Most loggers will only rarely be able to name a fossil to species level, but the broad grouping should be given where possible—for example, bivalve, ammonite. Graphic symbols exist for the more common types (Appendix 1). Samples may be taken for expert identification if required. The arrangement of fossils in the core can be indicated on the graphic log and detailed in the description. It is important to note whether the fossils are obviously in life position, or disordered, abraded or broken.

Because of the limited diameter of cores, the recovery of good-quality macrofossils tends to be poor even in a formation that in outcrop would be considered as richly fossiliferous. Because of this, the disciplines of micropalaeontology and palynology have been widely used in core studies, especially in the oil industry. Microfossils, pollen and spores are widely distributed in sediments, and may occur in substantial quantities even in small core chips. Microfossils occur in a variety of sediments, and are especially abundant in most limestones. Spores and pollen, however, are best preserved in poorly oxygenated sediments, and might be sought in dark shales and silts, in addition of course to coals. If fossil evidence is important to core interpretation, the logger should be aware of and look out for those lithologies in which microscopic material is most prevalent.

4.3.10 Accessories

These are defined as minerals, or other rock components, that are present in a rock in insufficient quantities to affect rock classification. As such they may not be of great significance, but once identified, and for the sake of completeness, they should be recorded except on small-scale logs. Of course, the significance of a mineral's presence cannot be judged just by its contribution to rock volume. This is obviously the case where one is prospecting for precious metals, when the mineral in question may never reach 1% of the rock. Similarly, a mineral such as glauconite, despite occurring in small amounts, may provide a major clue to depositional environment, and others may form correlatable horizons. The logger needs to decide when an accessory mineral is of significance, in which case it may be highlighted by including its symbol in the graphic lithological column.

Other accessories that might be encountered in a core include miscellaneous rock fragments, organic matter and bitumen.

4.3.11 'Sedimentary' structures

This heading here covers all non-tectonic structures, so bioturbation and even layering in igneous cumulates are included.

Structures occurring in core are essentially visual patterns, and often the simplest and most effective method of recording these structures is to draw them to scale on the graphic lithological column (Fig. 4.11). On a detailed log this column might become overcrowded with both lithological and structural data. In this situation, these datasets can be drawn on two adjacent columns (Fig. 4.11).

There are numerous symbols defined to cover various sedimentary structures (Appendix 1), but they are far less effective on the log than a good sketch. A symbol for 'tabular cross-bedding', for example, may encompass a variety of structures, whereas attempts to describe the structure more precisely in the text will demonstrate the truth of the adage 'a picture is worth a thousand words'.

Some structures cannot be represented clearly in the two-tone line drawings possible on most logs, and there are also difficulties in drawing intricate structures on small-scale logs. In these cases it will be necessary to resort to symbols. It is still possible, though, to draw the larger, clearer structures freehand, but intersperse them with symbols where necessary (Fig. 4.11).

One aspect of sedimentary structure that is sometimes represented quantitatively on a log is thickness of bedding. This may be of value in sedimentological interpretation and in engineering studies, and may be plotted in a separate column close to the graphic lithology column (Fig. 4.11). Knill *et al.* (1970) have recommended a 'classification of bedding spacing for engineering purposes' that was originally developed by Binnie and Partners and is now included in British Standard 5930:1999:

Description	*Bedding plane spacing*
Very thickly bedded	Greater than 2 m
Thickly bedded	0.6–2 m
Medium bedded	0.2–0.6 m
Thinly bedded	60 mm – 0.2 m
Very thinly bedded	20–60 mm
Laminated	6–20 mm
Thinly laminated	Less than 6 mm

The terms 'laminated' and 'thinly laminated' in this classification are reserved for sedimentary rocks, and are replaced by 'narrow' and 'very narrow' for igneous and metamorphic rocks. The formal classification, sensibly, allows some flexibility in the language, so that 'SANDSTONE, thick beds' is the same as a 'thickly bedded sandstone', and in igneous and metamorphic rocks the term 'bed' may be inappropriate: a 'very thinly flow-banded diorite' falls within the same classification system.

When representing sedimentary structures pictorially on a log, a decision needs to be made as to whether to draw them as they appear in core, incorporating any dips resulting from tectonic structure or well deviation, or to attempt to represent the structures as they would appear with any post-depositional dip removed. The decision would depend in part on the purpose of the log. Core logs produced, for example, as part of structural, geotechnical or mineral exploration work would generally incorporate structural dip.

Where the purpose is to interpret depositional environments, it may be preferable to attempt to draw the log with this dip 'removed', although sometimes there is insufficient information to do so unambiguously. Quite often, however, a sedimentary succession will contain mudstone or claystone intervals with bedding that would have been horizontal at the time of deposition, since they represent deposition from suspension of sediments that are too fine-grained to sustain bedforms with an appreciable sedimentary dip. Any dip on these beds must therefore result either from post-depositional tilting or from well deviation. This would greatly assist any attempts to restore sedimentary structures to their dip angles at the time of deposition. Other structures indicate the vertical at the time of deposition, such as vertical burrows, desiccation cracks or fluid-escape structures. If the bedding dip is removed on the log in this way, the fact should be recorded.

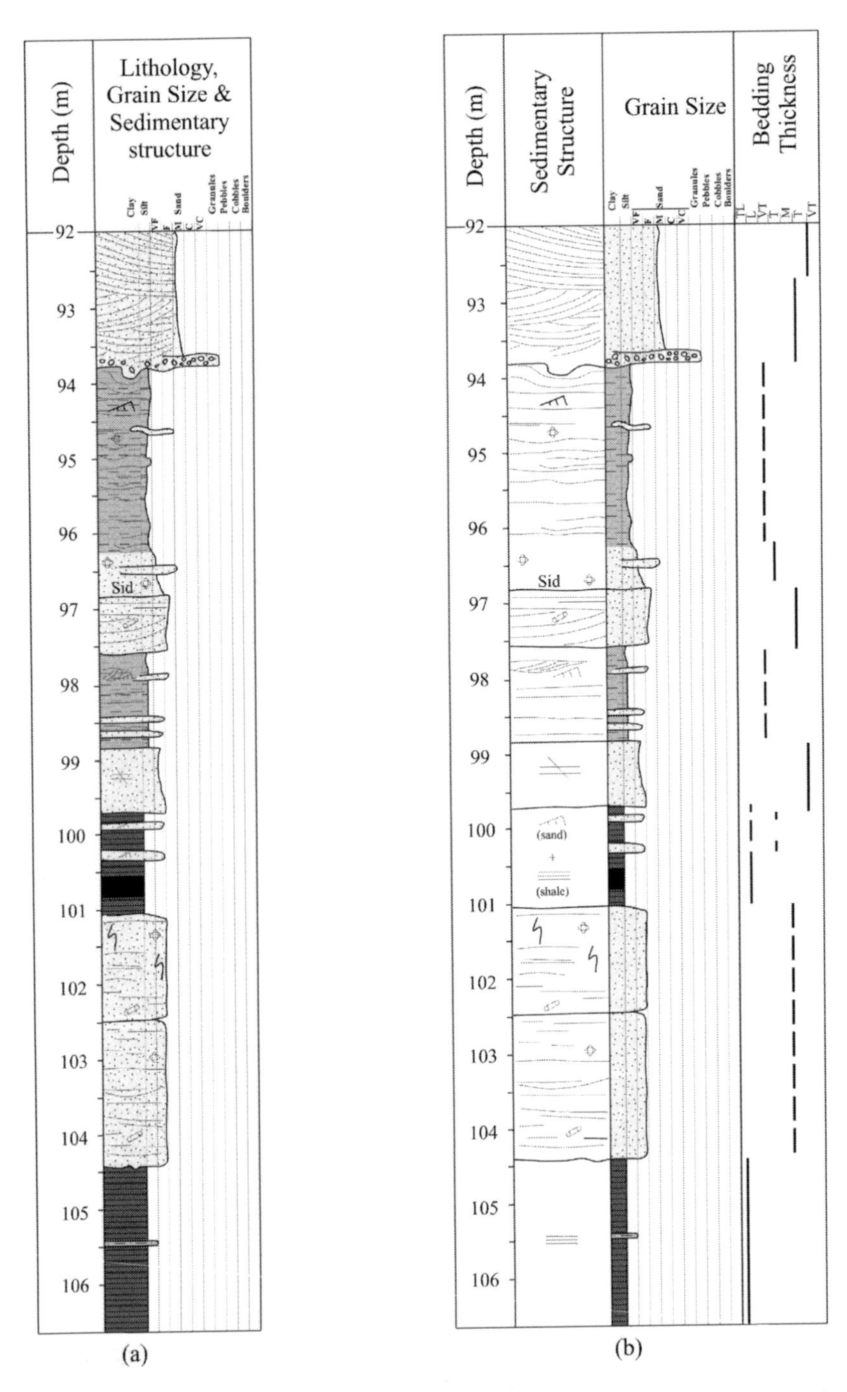

Fig. 4.11 *Lithology and sedimentary structures may be drawn (a) on the same graphic log, or (b) in separate columns. Note the combination of freehand drawings of structures (the cross-bedding at the top), and the use of conventional symbols. A plot of bedding thickness is also shown.*

Where any post-depositional dip is represented on a core log, the dips should ideally be drawn exactly as they appear in the core. If, for example, originally horizontal bedding cuts the core at 45° (either because the well was deviated, or because the formation has a tectonic dip), it should be drawn with this orientation on the log. However, the ideal of logging structures exactly as they appear is often not possible, or even advisable, in practice. If logging is taking place from slabbed core, and slabbing has been carried out in an arbitrary orientation, the dip appearing on the cut face will be variable, between zero and the full structural dip. This situation is unsatisfactory, since it could give a false impression of varying dip or cross-bedding. If the outside of the core is clean, it may be possible to log from this in a consistent direction parallel to greatest dip, ignoring the cut face. If however, the surface of the core is obscured by mud or coring marks, then there is little choice but to log this apparently varying dip, although a marginal note on the log explaining the reason could usefully be added.

One solution is to ensure that core is always slabbed in a direction parallel to the greatest observed dip, although, as explained in Section 3.3.2, this is not always possible.

Even where slabbing has been undertaken consistently, the logger may have access to only half of the resulting core, and the chances are that it is not the same half all the way down. In other words, when the core is laid out, some sections will dip to the left, and some to the right, even though the dip downhole may have been uniform. The impression from the log, if this were drawn realistically, would be of reversing dips, or perhaps some type of cross-bedding. The best solution here, although not ideal, is for the logger to compile the log with a consistent dip in one direction. Some slabs will need to be depicted in mirror image. This can be justified, since it should avoid confusion, and in any case the slabbed face of the 'missing' section of half core would be a virtual mirror image of the one logged.

The above strategy of logging consistent dips carries a slight danger. It is possible that a core of cross-bedded sandstone, for example, may genuinely display dips in alternate directions, and a zealous logger, desiring uniformity, may disguise this by logging only undirectionally. However, so long as the logger works intelligently, this danger is only slight, since the opposing dips are likely, at least occasionally, to occur together in an intact section of core. The cross-bedded nature of the succession will then be exposed. It is statistically highly improbable that cross-bedding truncations in a sandstone succession would consistently have been obscured by breaks in the core that are too disturbed to reconstruct. Once the logger has recognized the true nature of the bedding, the log must be drawn so as best to reflect this. It may still not be clear whether, for example, dips to the 'left' and 'right' should occur in equal numbers, or whether one direction is more common than the other, but a careful examination of the core should go at least some way towards answering this.

Freehand sketching of bedding and other structures should preserve the angular relationships between the various features wherever possible, especially on larger-scale logs. Sometimes it is important to know the precise angles of stratification, fractures etc. to the core length and to each other. They can be measured with a protractor, and recorded numerically either in a column on the log set aside for this purpose, or by annotating the graphic lithological column. Note that, for the graphic lithological column to be drawn to scale with the angular relationships preserved, the width of the column must also be drawn to scale. This may be impracticable at smaller scales.

Ultimately, of course, it has to be recognized that information on the relative bedding dip of an unorientated core is ambiguous. If one is fortunate in having an orientated core, then the structures can be realigned to their position in the formation. This is one occasion where core can usefully be slabbed other than parallel to greatest dip (Section 3.3.2). By carefully slabbing the core in a (say) north–south direction, the features can be logged as they would appear in a north–south-orientated outcrop (Fig. 4.12). In many cases, though, this may not be so useful as logging the core in the normal fashion, showing maximum dip, but with a note beside each section indicating the orientation of that maximum dip direction with respect to true north (Fig 4.12).

4.3.12 Diagenetic features

These include all those mineralogical changes resulting from low-temperature (non-metamorphic) processes acting on a rock body since initial sedimentation or cooling. One of these, cementation, has already been discussed, but minerals may be precipitated without acting as cement, and diagenesis also causes alteration or dissolution of minerals. Detailed diagenetic descriptions require laboratory examination of thin sections and other types of analysis, but a hand lens or binocular microscope is sufficient for recognizing the most important aspects. It is normally quite easy to spot, for example, euhedral quartz overgrowths, grains that have altered to a powdery clay, and small vugs resulting from dissolution of one or more minerals.

4.3.13 Weathering

Weathering is strictly a form of diagenesis, but in near-surface cores (or deep cores below unconformities) it can be the most significant feature of a rock, especially in engineering investigations, and should be carefully described. Where it is important to define the degree of weathering, there are several classifications that may be used. That defined by Knill *et al.* (1970) is as follows:

- IA Fresh (F): no visible sign of weathering.
- IB Faintly weathered (FW): weathering limited to the surface of major discontinuities.
- II Slightly weathered (SW): penetrative weathering developed on open discontinuity surfaces, but only slight weathering of rock material.
- III Moderately weathered (MW): weathering extends throughout the rock mass but the rock material is not friable.
- IV Highly weathered (HW): weathering extends throughout rock mass and the rock material is partly friable.
- V Completely weathered (CW): rock is wholly decomposed and in a friable condition, but the rock texture and structure are preserved.
- VI Residual soil (RS): a soil material with the original texture, structure and mineralogy of the rock completely destroyed.

The British Standard *Code of Practice for Site Investigations* (BS 5930:1999) acknowledges that forms of weathering in different rock types are so diverse that formal classification may often not be appropriate, but it does provide guidelines for varying circumstances. Where required, the degree of weathering may either be incorporated in the written description, or be plotted with depth on a scale following, for example, that of Knill *et al.*

4.3.14 'Tectonic' structures

These are fractures and faults, in addition to rock cleavage, gneissose banding, schistosity and tectonic folding. As with sedimentary structures they are best recorded by sketching, either in the graphic lithological column or in a specially designated column alongside it. Precise detail regarding angles between structures, and relative to the core length, can be added numerically where required (Figs 4.10, 4.12).

In engineering investigations fractures will be among the most significant properties in a core. Several methods are conventionally used to measure the degree of fracturing. The 'solid core recovery' (SCR), mentioned in Section 3.2.4, is the percentage of the cored interval recovered as solid (full-diameter) core. When compared with the total core recovery (TCR), it gives a simple measure of the amount of fragmentation in the core.

A 'fracture log' provides a measure of the number of natural fractures occurring over an arbitrary length of core (Fig. 4.13). The 'fracture index' (FI) is the number of fractures per metre. It requires fractures resulting from damage to the core, either during coring or subsequently, to be recognized and discounted (Section 4.3.18). Other more sophisticated measures of the fracture state of core are the rock quality designation (RQD; Deere *et al.*, 1967) and the stability index (Ege, 1968). The RQD is the percentage of solid core recovered comprising pieces greater than 0.1 m (previously 4 inches) in length. The stability index comprises the sum of a number of rock properties, and is defined as follows:

> Stability index = 0.1 × core loss (length drilled less total core recovery) + number of fractures per foot + 0.1 × broken core (core less than 7.5 cm in length) + weathering (graded 1 to 4 from unweathered to severely weathered) + hardness (graded 1 to 4 from very hard to incompetent).

The stability index can then be graded on a rock-quality scale from 1 to 10, with 1 referring to incompetent rock (index greater than 18) and 10 to good rock (index less than 8). Note that the stability index takes account of other rock properties besides simple fracturing.

One or more of these indices of rock quality can be determined and plotted graphically on a column of the core log.

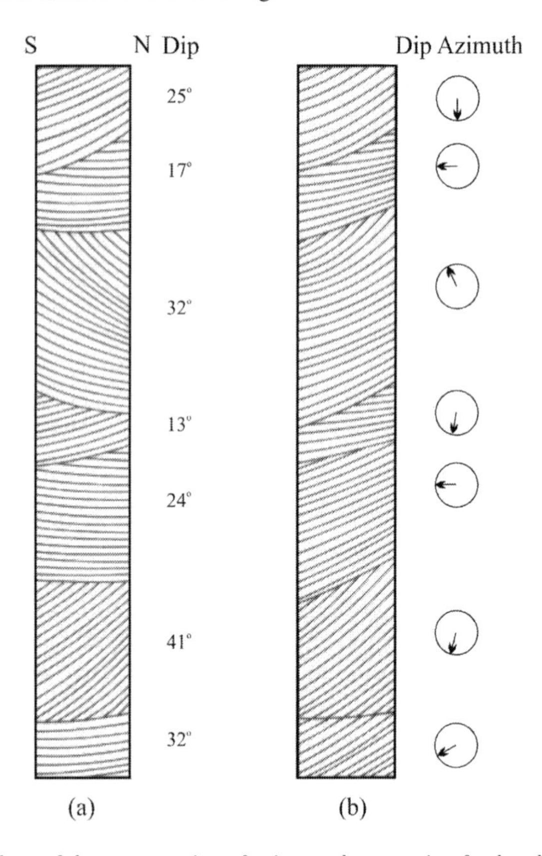

Fig. 4.12 *Two logs of the same section of orientated core, using freehand drawing of sedimentary structures. In (a) the structures are depicted as they would appear in a north–south-oriented outcrop. The angles marked are the maximum dips on each cross-bedded set, which would normally be in a direction out of the plane of section. In (b), each cross-bedded set is logged as it would appear if slabbed parallel to its maximum dip direction. The arrows next to each set indicate the direction of that dip with respect to the north (top).*

Fractures may be open or closed, and may or may not be mineralized. This information should be recorded, together with details of any apparent movement along the fractures. Faults may be associated with drag on one or both margins. Great care needs to be taken to distinguish original fracturing and faulting from those caused by the coring process (Section 2.9). Cementation of a fracture is a good guide to its original nature, so long as the 'cement' is not dried drilling mud. Original uncemented fractures may nonetheless display mineralization of the fracture surface or, especially in near-surface cores, differential weathering. A fracture that can be fitted together perfectly is likely to be an artefact from the later stages of core recovery, or from later mishandling, since drilling mud will enter along original subsurface fractures during coring, preventing a subsequent good match of the two sides. Fractures generated by the torque on the core barrel may display a helical geometry, which is unlikely to have occurred naturally. Apparent slickensides in core are sometimes a result of differential movement on two sides of a fracture (original or otherwise) during coring. Unless associated with mineralization or other convincing evidence of their original nature, consideration should be given to the possibility that they are artefacts.

All the guidelines for logging dips in cores consistently, so as to avoid giving a false impression of alternating dips (Section 4.3.11), should also be adhered to when dealing with tectonic structures.

4.3.15 Veins and mineralization

Veins and mineralization could mostly be included under categories already listed, such as mineralized fractures, cementation and other diagenetic features. However, these sections treat mineralization as a relatively minor aspect of the core, whereas to the economic geologist they may be of the greatest importance. Although the principles of logging will remain the same, the mineralization and its nature will need to be emphasized. Mineral species present can be listed in a separate column, and sketches illustrating different phases of mineral growth and cross-cutting relationships can be used. Estimates may need to be recorded of the proportion they represent of the total core volume. Samples may need to be taken for laboratory identification and analysis, and the position in the core from which the sample has been removed must be carefully recorded.

4.3.16 Porosity and permeability

These are important in both hydrocarbon and groundwater investigations, in addition to surveys of potential waste disposal sites. Accurate measurements require laboratory analysis, but with practice a reasonable estimate can usually be made from inspection of core samples. Various methods also exist for determining permeability and other parameters based on the flow of fluids within the well (either naturally, or tested under elevated or reduced pressures), although these are not considered here.

On laboratory analysis even the least porous lithologies, such as mudstones and granites, are usually found to have a porosity of several per cent. This may partly be due, however, to microfracturing resulting from the pressure release involved in retrieving the core from depth. Such porosity is best described as negligible rather than zero. For other samples, porosity can be divided by hand specimen examination into ‘poor’, ‘moderate’ and ‘good’. Poor-porosity samples are those in which pores are evident, but are very limited in size or abundance. They may correspond to actual porosity of about 5–10%. Moderate porosity indicates a good visible pore structure, but limited perhaps by clay matrix or cement. Actual porosity values may be 10–20%. Good porosity indicates a well-developed system of pores, which may display some cement or matrix, but enough only to hold the rock together without greatly reducing total pore volume. Laboratory analysis may indicate 20–30% porosity. Exceptionally porous sands, which may exceed 30% in measured porosity, can be designated ‘very porous’. Rocks of this type will have a very open structure, and if not totally unconsolidated will probably have a tendency to disaggregate.

Permeability is more difficult to judge from hand specimens, but with experience a reasonable estimate can be made. A useful qualitative field test is to drop a little water on a piece of clean dry core, and observed the speed at which it is absorbed. With negligible permeability the drop will form a bead on the surface, whereas with very good permeability the water will sink immediately into the rock. Between these two extremes the water will take a variable time to be absorbed, ranging up to several minutes in barely permeable lithologies. This speed of absorption depends on numerous factors, such as mineralogy and pore structure, but it can give a useful indication of permeability. Unfortunately, this test does not work with oil-stained cores, in which even with permeable rocks the water often forms a bead on the surface rather than being absorbed.

An even simpler test for permeability is to place the hand specimen firmly against the lips and blow. If the rock has negligible permeability, it will appear impossible to blow through. With low-permeability samples it is usually possible to detect some leakage of air through the rock, whereas a steady flow, despite some obvious resistance, may be classed as moderate permeability. A good-permeability rock can easily be blown through, and with very good permeability many of the rock grains will probably crumble off into the mouth—beware!

The porosity and permeability designations of a particular lithology—moderate, poor, etc.—will often coincide, but this is not necessarily the case. A vesicular basalt, for example, may have a very good porosity but negligible permeability, and a quartz-cemented arkose in which feldspars have dissolved since cementation may display a similar if less-extreme disparity. It is therefore unsatisfactory to assume that good porosity will imply good permeability, and a separate test needs to be applied for each.

In addition to the amount of porosity, the porosity type may need to be recorded. Detailed porosity classifications exist, and will be found by reference to the bibliography. For most purposes it is sufficient to record whether the porosity occurs between the main rock-forming particles—intergranular or intercrystalline—or by dissolution of the particles—intragranular or intracrystalline. If the latter is the case, and dissolution of a particular type of particle such as a feldspar grain or shell fragment is the cause, this should also be recorded. Vuggy porosity is generally caused by dissolution that produces voids the same size as or larger than the average size of rock-forming particles in a particular lithology.

Fracture porosity occurs in open fractures. Values of fracture porosity are only meaningful, however, over a relatively large volume of rock, and require careful measurements of the volume of porosity in open fractures over a measured length of core. In igneous rocks, vesicular porosity may occur.

4.3.17 Oil shows

These will be most commonly encountered in wells drilled specifically for hydrocarbon exploration, but they can occur in any sediment, and may frequently be found in thermally mature coal-bearing sequences.

Oil is most easily recognized as a sticky brown substance impregnating the core. At the wellsite the pressure release may cause oil to bubble out and bleed from the core as it is removed from the core barrel. The colour of the oil should be noted. Lighter oils do not have this sticky brown character, and may be recognized by their odour rather than their appearance. This odour should also be noted, together with an indication of its strength. A faint odour may be detected only on a fresh surface, or after a core piece has been kept in a closed bottle for some minutes.

Petroleum geologists have a range of tests for oil in cores (see, for example, the AAPG *Sample Examination Manual* listed in the Bibliography). The commonest is examination under ultraviolet light, which causes the oil to fluoresce. The colour and distribution of fluorescence should be noted.

Care needs to be taken, as some minerals may fluoresce, notably anhydrite, fluorite and sometimes calcite. The test is made more effective and reliable by observing a small core piece under ultraviolet light while it is immersed in an organic solvent such as chlorothene. This is termed a 'cut fluorescence' test, and if hydrocarbons are present, fluorescent 'streamers' will be emitted into the solvent. The colour and intensity of these streamers should be noted.

It is useful to record the distribution of oil in a core, and relate it to variable lithology and structure. It will often be found that differential oil staining highlights sedimentary and tectonic structures. It is important not to assume, however, that the most heavily oil-stained lithologies are the best reservoir lithologies. The degassing effect of bringing a core to the surface will often force fluid out of the good-quality permeable sands, leaving only a light residual oil staining. In poorer-quality sands, however, the process will force the oil out less rapidly and uniformly. The lighter hydrocarbon fractions will tend to evaporate from the remaining oil, leaving the characteristic and very noticeable sticky, dark brown staining. Although this is certainly indicative of porosity in the rock, it may also indicate a poor or heterogeneous permeability, which will lead to oil production problems.

The presence of a distinct 'tarry' zone at a specific horizon sometimes indicates the location of a (current or 'fossil') oil–water contact within a hydrocarbon reservoir, and is especially worth recording.

4.3.18 Core damage

It is extremely rare for a single intact cylinder of rock to emerge from a core barrel, and as noted in Section 2.9 there are numerous types of core damage that the logger needs to recognize. The core log should record, as accurately as possible, the likely state of the core before it was cut from the formation, so the purpose of recognizing types of damage is generally in order that they may be discounted and omitted from the log. There are, however, types of core damage that can so seriously affect the quality of the log that their occurrence in the core should be recorded. In addition, core damage can indicate some characteristic of the formation (such as a tendency to fracture) that would otherwise not be apparent. This should also be noted on the log.

Frequent breaks along the core are to be expected. These result from (among other things) fracturing due to the torque on the core during drilling, natural fracture breaks along lines of weakness in the formation, breaks where drill-string connections were made during a coring programme, and damage incurred in removing the core from the barrel at the wellsite. Distinctive 'chevron fractures' can occur along the length of a loosely consolidated core recovered in a liner, resulting from flexing of the liner as it is removed from the barrel at the wellsite – the 'chevron' pattern is in fact the two-dimensional appearance of stacked conical fractures. Damaged core can usually be more or less reconstructed, but in places 'rubbly zones', comprising a disordered collection of rock fragments, occur. This type of core damage is frequently encountered at the base of a core, and less frequently at the top. Although the lithologies present can be examined, their order, and any structures they exhibit, cannot usually be reliably reconstructed. The log should be drawn up as accurately as possible, but the poor quality of the data over the rubbly zone should be clearly indicated. This is often drawn with a heavy zigzag line alongside the graphic lithological column (Fig. 4.13).

As noted in Section 4.3.3, rock fragments falling downhole immediately before and between separate coring runs may result in a disordered accumulation at the top of the core. These may or may not correspond to the lithologies occurring immediately above and below, and may be a mixture of nearly *in-situ* and exotic fragments. If the fragments are identical to immediately adjacent lithologies, it will probably be impossible to demonstrate whether or not they were recovered *in situ*. Rock fragments occurring at the top of the core that seem out of place should generally be logged, just in case they have a significance unclear to the geologist at the time. The doubts as to their true affinity should, however, be clearly expressed on the log.

A rubbly zone occurring in the middle of a core needs to be investigated more closely. It may result from coring difficulties not associated with the nature of the formation. This should be apparent to the wellsite geologist, but whether it can be checked later will depend on the availability and completeness of drilling records. Alternatively, a rubbly zone may result from an interval of weakness or fracturing in the formation, or may even be a poorly consolidated sedimentary or tectonic breccia that has been recovered intact but misinterpreted as resulting from core damage. In any event it needs to be recorded in the log, as its presence can influence subsequent interpretation.

Missing sections are common in cores. They are usually detected at wellsite when, having completed a coring run with a 60-foot barrel, say, only 57 feet are found to have been recovered. Since such losses usually occur from the bottom of the barrel as it is lifted to the surface, the wellsite geologist will, in the absence of evidence to the contrary, assume this to be the case, and mark depths on the core accordingly (Section 3.2.4). Of course, if the missing 3 feet has somehow, unknown to the wellsite geologist, been lost from the top or middle of the core, depth marks on the remainder of the core below the true position of the missing section will inadvertently be marked 3 feet too shallow.

Therefore, when logging a core in which losses have been recorded at the wellsite, the logger should be aware that the missing section may not be exactly where indicated by the depths marked on the core. It would be rare for a logger to find unequivocal evidence from the core alone to amend the depths, but data may be available to which the wellsite geologist did not have access.

Electric wireline logs, for example, may indicate the true relative depths of thin beds of shale or coal. If the logger has the wireline logs for comparison with the core while logging, the correct position of the missing section may be spotted. It must be emphasised that the depths marked on the core should not be altered, unless the logger is positive that this will not cause confusion to anyone else who has looked at it, or who uses data or samples already collected from it. The safest course is never to amend marks on core or core boxes.

However, the log can be drawn up according to the revised core depths, with a clear note as to the justification for doing so (for example, 'core between 929 ft and 947 ft shifted down 3 ft from depths marked on core, to match shale beds at 936 ft and 942 ft with pronounced spikes on gamma-ray log'). It may also be helpful to leave a signed and dated note in the relevant core box detailing the suggested depth changes.

A further note of caution is needed. Just as the depths marked on core should not be amended, so samples taken from core should always be labelled with the depth marked on the core, even if this does not correspond to the logged depth. This is because a well may be logged numerous times, both by different geologists looking at the same core, and by wireline techniques. There is thus any number of 'log depths' but only one 'core depth' (defined as that originally measured by the driller or wellsite geologist). The convention of relating core samples to core depth is straightforward and unambiguous, so long as the geologist remembers to correct core depth to whichever particular log depth may be in use at the time. This system also enables samples to be reunited with the correct core box, and additional samples to be taken at a stated depth, without risk of confusion.

In addition to the possibility of a missing section being relocated on logging, the logger may also find that a missing section is not missing after all! This happens when the bottom section of core slips from the barrel as it is being withdrawn and remains at, or falls down to, the bottom of the well. The wellsite geologist will record this, correctly, as a missing section at the base of the core. However, if the core barrel is run into the well to take another core, the 'missing section' may be the first core to enter the barrel, either intact or, more probably, as a rubbly unit. It will thus be recorded at the surface as the top section of the second core. Once more, it is possible that the watchful logger will spot this, even though it may have escaped the attention of the wellsite geologist.

Marked narrowing or necking of the core can occur. If this occurs as an irregular or sporadic loss of some of the outer portions of the core, it may be due to weak or soft core losing material into the mud system. Although drilling mud flows down between the inner and outer core barrels, in some designs of barrel it washes past the lowermost few centimetres of the core as it is being drilled, and may cause erosion of the core in soft or friable formations. Also, some types of core catcher have sprung prongs that retract as the core enters the barrel during drilling, but which sometimes erode the outer margin of the core before retracting fully. This may produce a cone-shaped termination at the top of a cored interval, or below some natural break within a cored sequence. The concentric or spiral marks of the core catcher prongs on the core will usually identify the cause in this instance.

If the narrowing of the core seems to relate to lithological properties rather than purely to the operational method, it may usefully be recorded on the log.

It will sometimes become apparent to the logger that core pieces have been placed in the core boxes out of sequence or upside down. This may be indicated by two out-of-sequence pieces that fit neatly together, or by a sedimentary or other structure that is unambiguously upside down. The principle holds once again that wellsite core markings should be regarded as untouchable. Even if the depth or way up marked on the core is obviously incorrect, the core markings should not be amended, nor the pieces moved. Instead, a note may once again be left in the core box drawing the attention of future viewers to the mistake. The log should be drawn up with the core pieces in their correct position and orientation (that is, not as represented in the core box), but a comment on the log should make clear that this is the case.

If mixing of the core has occurred since it left the wellsite, so that the marked core depths are themselves displaced, or way-up arrows are incorrectly orientated, the solution is simpler. It is then permissible for the logger to restore the core in the box to its original position, and log accordingly. The only proviso is that care should be taken that this does not cause difficulties for anyone who has already used the core, for instance for sampling, and who may not have noticed the mix-up. If this is likely, it may still be best to leave the core in its disordered state, but with another slip of paper explaining the situation. Alternatively, of course, it may be possible to contact the other core user, giving them details of your rearrangement of the core. This is a good example of the value of leaving a note in the core box of samples taken, together with the name and location of the sampler, and the date (Section 3.2.5).

4.3.19 Samples

The geologist logging core will often also be taking samples. Notes concerning core sampling may be found in Sections 3.2.5 and 5.1. Although logging and sampling are two separate functions, they are complementary: it is usually the information derived from logging that allows the geologist to pick sampling points, and samples will usually themselves be analysed to yield data that can later be added to the log.

In addition to labelling samples with depth and well, it is therefore useful also to mark the sampling point on the log, together with a note detailing the type of analysis to be carried out. The simplest method is to draw annotated arrows immediately adjacent to the lithological column, indicating the precise sampling depth (Fig. 4.4). Occasionally a depth alone is insufficient, as the sample may be taken from beside some lithological boundary or other structure that cuts the core obliquely, so that a single depth could represent two or more lithologies. The arrow would in this case point to the specific point on the graphic lithological column from which the sample was taken (Fig. 4.14). If the scale of the log is insufficient to do this unambiguously, then a larger-scale sketch in the comments column might be used for clarification.

In addition to details of samples taken during logging, information regarding samples taken by others at an earlier date may need to be recorded on the log. If, for example, whole sections of core have been removed, logging will not be possible over the affected intervals, and the log must be marked accordingly (Fig. 4.14).

If good practice has been followed, a note in the core box should give precise depth details of the samples removed (and perhaps even a brief lithological description). Sometimes it will be possible to examine these samples at a later date, and fill in the missing sections of the log.

4.4 Trends, sequences and cycles

The main purpose of constructing a log is usually to produce a record of the chief features of a core. Despite the geologist's best efforts, there are inevitably numerous details that are omitted or recorded imprecisely in producing a log: hence the value of storing core permanently wherever possible. However, other information from the core becomes clear only on completion of a log. In this respect, a core log should be thought of not simply as a summary record, but also as an essential tool in geological interpretation. To take a simple example, the interpretation of depositional environments often depends heavily on the recognition of certain vertical grain-size trends. If one inspects core without logging, it is difficult to assess grain-size variations in more than a general sense. This is especially true if, for example, 100 m of core is stored in 1-m boxes, and there is only space to view one box at a time. A broad upward fining, for instance, might be recognized, but interpretation may depend critically on the rate of fining over certain intervals, or on more subtle variations superimposed on the overall trend. On a core log that records grain size graphically (Section 4.2.4), so long as grain size has been carefully measured along the length of a core, both the overall trend and any more subtle variations should immediately become obvious.

The grain size curve is just an objective record of measured data. However, the geologist will often want to highlight the existence of certain trends, such as the presence of a fining-upward sequence. This is often shown by drawing an arrow alongside the grain size column, pointing in the direction of fining.

Sometimes, a large-scale upward coarsening 'trend' (for instance) may comprise a number of individual fining upward 'cycles' (as when a series of stacked turbiditic 'Bouma' cycles occurs in a prograding submarine fan succession). In this case, one long downward-pointing arrow on the log may be placed alongside a succession of upward-pointing arrows (Fig. 4.15).

These grain-size trend arrows are a less objective indicator than the simple grain-size graph, since they indicate those sequences that the geologist considers are, or might be, significant. Nonetheless, they can be very valuable in highlighting features that might not otherwise be immediately obvious. In addition to their use for highlighting grain-size trends, each of a sequence of fine-grained lava flows, for example, may display a slightly reddened

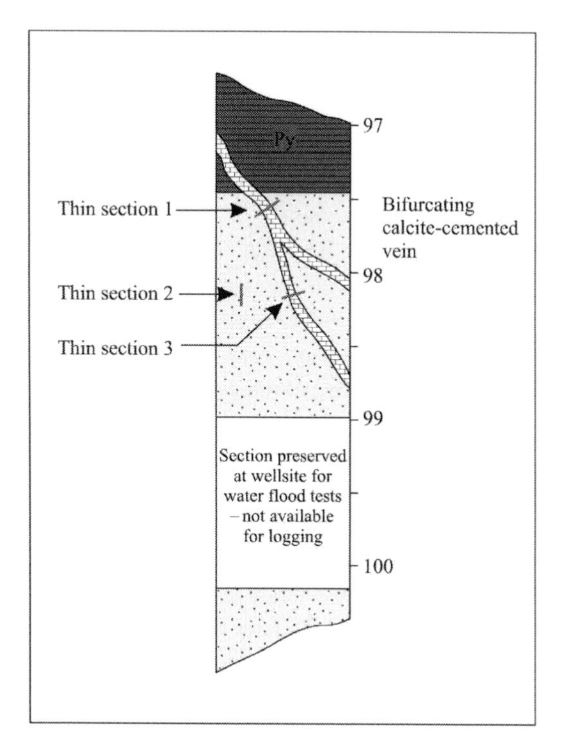

Fig. 4.14 *On a large-scale log, the precise location of samples taken can be marked on the graphic lithological column. This is often more informative than a simple depth.*

weathering zone at the top. These may be evident to the logger, but the record of the slight colour change would be buried amongst the text of the log, and may not be noticed by subsequent users of the log. Suitably labelled trend arrows would clarify the position, and emphasize an important but subtle feature of the succession.

Trend arrows and the like can be used wherever the geologist wishes to highlight some depth-dependent variable of the rock. In addition to grain-size variation and volcanic cycles, they can indicate changes in the degree of fracturing, mineralogical composition, angularity of clasts, colour, or anything else that the geologist thinks may help in the interpretation of the core. The arrows may be drawn in a column especially reserved for 'trends' (Fig. 4.15), or may be situated in any other convenient position on the log, such as beside the lithology column, or the written descriptions. It is, of course, important to ensure that the arrows are properly labelled, so that it is quite clear to what they refer.

4.5 Completion of preliminary logs

The preliminary log refers here to the log drawn up by the geologist on site, or in the core store. It may be tidied up back in the office, but it contains only data or interpretations obtained during logging of the core. There may be no intention to develop the log beyond this point, but alternatively it may have other data, interpretations or laboratory analyses added at a later stage, and may be redrafted at a different scale. The completed log, which eventually is filed for long-term reference, is here denoted the 'final log'. In many cases the preliminary log and the final log will be one and the same.

On completion of logging, the first task, while the core is still at hand, is to run through the log to ensure that all the necessary data have been entered. It is very easy to omit data accidentally from the log on the initial logging run, and missing information is difficult to obtain on returning to the office. If one has been strictly methodical in logging, moving from one column to another in an ordered sequence, the likelihood of omissions here is much reduced.

Before leaving the core, one should make sure that the log is in a sufficiently comprehensible state that, if necessary, a colleague could use it without further explanation. It is important to ensure that everything is clear and legible. The intention may be to neaten the scribbled writing that afternoon at one's desk, while it is freshly in mind. Unfortunately it is too easy to be waylaid, and to discover on returning to the log a week later that details cannot be read and have escaped the memory. Ensure that anything that is not perfectly standard is clearly explained. If, for example, the core is full of lenticular bedding, and because one has not come across this before a symbol has been created, a full explanation of the symbol should be included on the log.

The resulting log is a valuable document, and should be treated with care. In many cases it will be heavily used during ensuing project work, and thought should be given to making a copy for day-to-day use, and filing the original. It may ultimately be drafted professionally, but this will not usually occur early in a project when amendments and additions are likely. During the progress of a project new data and comments will often be added to a log, and it is best to do this on a good copy. Data derived from initial core viewing can then be distinguished easily from those subsequently added.

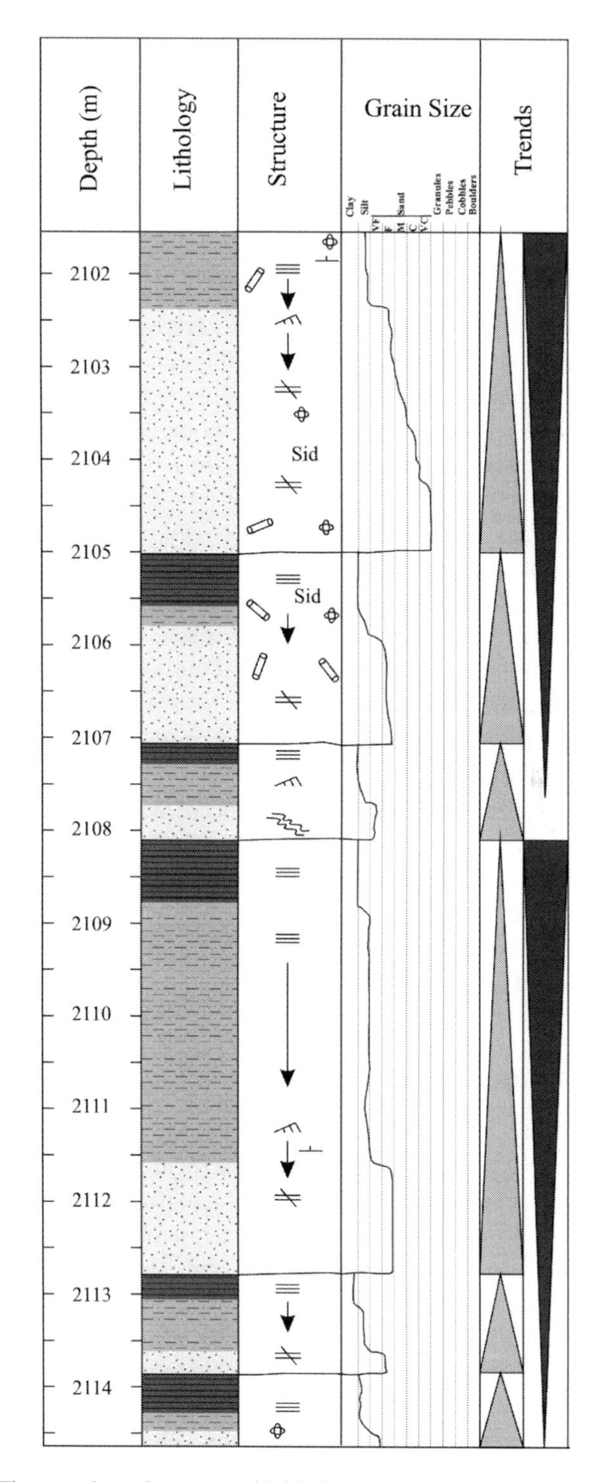

Fig. 4.15 *The use of trend arrows to highlight grain-size variations at different scales.*

5 Core analysis and testing

There is an indefinite number of ways in which core can be analysed, and it is not the intention of this chapter to detail any or all of them. Rather, the aim is to outline some of the ways in which analysis of core samples differs from that of, say, outcrop samples, with particular emphasis on the pitfalls of using core. In addition to this, a few analytical techniques for which core is particularly suited will be highlighted.

Geologists who are unused to working on core will mainly be accustomed to using outcrop samples. The differences between the two need to be kept firmly in mind. Core can present the geologist with a continuous vertical succession through a formation, sited where it is needed, rather than depending on the vagaries of geomorphology (and sometimes the attitude of landowners) that control the location of suitable outcrops. If the core comes from a substantial depth, it will not have suffered weathering, and it may contain remnants of original reservoir fluids, unaffected by recent near-surface meteoric ingress. On the other hand, core is only of very limited width, and will not display the lateral variation that would be visible from surface outcrop. The lack of weathering in core is not wholly an advantage, since differential weathering can highlight structures that would not be apparent on freshly cut cored surfaces. However, core can recover soft or soluble lithologies that would be unlikely to form significant surface exposures in most climates.

5.1 Sampling technique

Sampling of core has to be undertaken with great care, for two main reasons. First, a core represents a unique resource. The same locality is rarely cored twice, so any core sample destroyed in analysis is unlikely to be replaced. Care therefore needs to be taken while sampling to ensure that the least possible disruption and loss are caused to the remaining core. These aspects of sampling are covered in Sections 3.2.5 and 3.3.2.

The other aspect of core sampling that needs great care is choosing the location and size of sample that will most faithfully reflect the conditions in the undisturbed formation. The act of drilling a well irrevocably alters the nature of the formation, both in the immediate vicinity of the well-bore and in any samples recovered. Penetration of the core by drilling mud, and the resultant displacement of formation fluids, is perhaps the single most significant alteration. In addition, there are the effects of the release of overburden pressure, causing potential deformation and microfracturing, and even the localized thermal alteration that may be caused by the drilling bit. The act of forcing the core into and out of the core barrel will cause disturbance to unconsolidated sediments. These are all factors that might easily be overlooked. Fortunately, the various common types of core damage listed in Section 2.9 will mostly be obvious when sampling, although in some cases, as with very rubbly core, they may make sampling difficult.

Maximum infiltration and disturbance to core inevitably occur around the margins. The depth that is affected varies, depending on the nature of the rock. A well-indurated, near-impermeable rock may display an almost imperceptible zone of alteration perhaps less than 1 mm in width. It is generally prudent to assume, however, that at least the outermost 1 cm will have been affected to some degree. Although electric log responses often indicate considerable depths of penetration of mud, and especially mud filtrate, into the borehole wall (tens of centimetres, or even metres), the penetration of mud into the core is usually considerably less. The bottomhole pressure of the mud system is usually designed to be slightly greater than the formation pressure, and this pressure differential will force the mud into the formation. The accumulation of a mud cake on the borehole wall creates an impermeable barrier that slows and ultimately stops this fluid transfer, although the pressure differential remains (Fig. 5.1). Although the original pressure of the fluid in the core is equivalent to that in the formation, the radially inward pressure from the mud into the core (in a horizontal plane), coupled with the virtual incompressibility of the formation fluid, causes the fluid pressure in the core to build up to that of the mud, with insignificant penetration of mud into the core. There will still be a pressure differential between the margins of the core adjacent to the mud and the fresh formation below the coring bit (Fig. 5.1), but the flow will tend to be concentrated downward along the core margins rather than radially into the core. There is still a good chance, especially with wider-diameter core (more than 5 cm say), that the centre of the core will have suffered only minimal fluid replacement.

Although the central part of a core is usually the least disturbed, it is not always either possible or sensible to sample the centre. A study of formation fluids would be virtually useless if it were undertaken on samples from the outer margin of a core, because of flushing by mud filtrate. However, a micropalaeontological analysis on the same samples would probably be valid. When choosing sampling technique, the analyses to be undertaken will usually govern one's decision.

The simplest sampling method, and that which causes least damage to the core, is to choose broken fragments of the appropriate size from the core box. The chief danger is that small, loose chips are not usually orientated, and easily

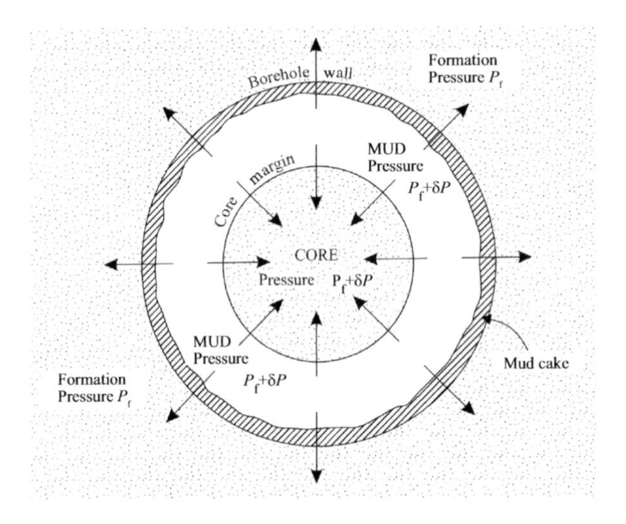

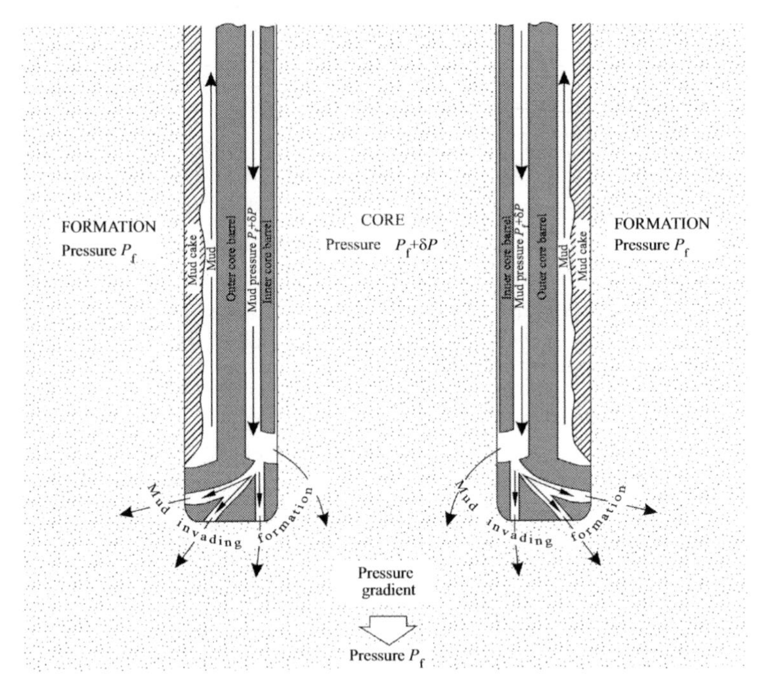

Fig. 5.1 *Simplified representation of pressure gradients and the movement of formation and drilling fluids in a permeable formation being penetrated by a coring bit and barrel.* P_f *is the original formation pressure, whereas the mud is (or should be) at a slightly higher pressure,* $P_f + \delta P$. *The pressure differential* δP *is responsible for fluid movements.*

become misplaced in the core boxes, owing to their rattling around during transport. This danger may be averted if the fragment sampled can be clearly identified as a chip off a larger and better labelled core piece. Otherwise, if sampling accuracy is important, loose chips are best avoided.

The use of a hammer to remove samples from larger sections of core is the commonest sampling method for indurated lithologies. It is usually easy to remove samples only from the outside margin of a core, especially if it has not been slabbed. A chisel may be helpful in sampling from the centre of a length of core, although it takes practice before this can be done without risking significant damage to the remaining core.

The simplest method of sampling the interior of a core while minimizing damage is to use a core plug drill (Section 3.3.2; Fig. 5.2). This is a heavy-duty bench drill fitted with a hollow cylindrical diamond bit that cuts out a small core or 'plug' from the rock sample. Plug diameters vary, but 25 mm is common (Fig. 5.3). Portable plug drills also exist, but they are rather unwieldy and need a source of water. Plug samples have the advantage of being accurately positioned and oriented in the core, and are of uniform size. Damage both to the sample plug and to the remaining core is minimized.

The preservation of friable or soft core by freezing has been described in Section 3.2.6. Frozen friable core is not usually easy to sample by hammering, as it tends to shatter rather than break cleanly. Drilling out core plugs is the simplest method of removing spot samples, although water, which is the most common coolant used with the core plug drill, is unsuitable for use with frozen core. The solution, as with slabbing of frozen core, is to use liquid nitrogen instead of water: this allows good-quality plugs to be taken, and the additional cooling of the sample, to a point well below 0°C, allows it to be handled briefly without thawing. In order for the plug to remain intact once it is thawed, it can be encased in a tightly fitting sleeve. Lead sleeves were once used, but Teflon or other plastic is now more common. This plugging of frozen core can be undertaken on either the unslabbed or the slabbed core. However, frozen core will often have been recovered and preserved within a sleeve of fibreglass, aluminium or similar material, and gaining access through this by slabbing or otherwise, even while frozen, may cause disturbance. An alternative is to use the core plug drill to cut plugs right through the sleeving, thereby causing the minimum of sample disturbance.

Problems are caused in a heterogeneous sleeved core where a particular lithology is required, as one is sampling 'blind'. The obvious solution is to keep plugging the core until a sample of the required lithology is obtained. A more elegant method is to use X-radiography or X-ray tomography (Section 5.2.3) to 'see' through the core sleeve and locate the required sampling points.

Sampling of preserved core pieces for the analysis of reservoir fluids can pose problems, since every effort must be made to prevent the fluids from being displaced or evaporating. As explained above, samples for reliable fluid analysis are more likely to be derived from the centre of the core. However, the use of a

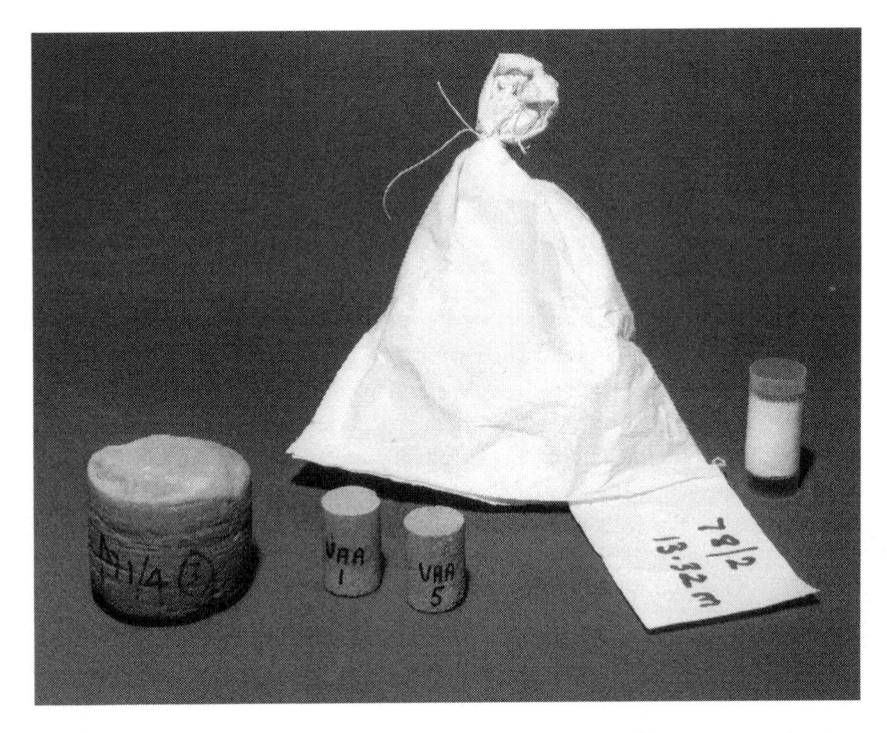

Fig. 5.3 *Various types of core sample. The bag contains a core chip, and on the left is a piece of full-diameter core. Full-diameter samples should be taken only where absolutely necessary, as they remove a complete section of core. The two cylinders are 25-mm plugs cut from a consolidated core, and on the right is a plug cut with a liquid nitrogen lubricant from frozen unconsolidated core, and encased in a plastic sleeve. Discs of wire mesh at each end allow porosity and permeability analysis to be undertaken.*

slabbing saw or core plug drill with associated drilling fluid is likely to contaminate the original reservoir fluid. Depending on the state of the core, a suitable piece may be removed from the centre with a hammer and, if necessary, a chisel, with every effort being made to minimize damage to the remaining core. It is important to take a single large sample rather than a handful of small fragments, since fluid will very rapidly evaporate from the latter. If the core has been frozen at wellsite so that the contained fluids are immobilized, the use of a core plug drill and liquid nitrogen coolant should allow undisturbed samples for fluid analysis to be taken from the centre of the core.

Even if the samples were not frozen at the wellsite, they may still be frozen later and plugged with liquid nitrogen. If, for example, a whole-core sample has been preserved in wax (Section 3.2.6), it can be frozen and a core plug cut through the wax coating. The same process can of course be applied to a core piece on which no preservation technique has been attempted, although the closeness with which the remaining core fluids correspond to the original formation fluids must in this case be held in considerable doubt.

The samples discussed so far, whether small or large, are all essentially 'spot' samples, used to analyse for a certain parameter at some point within the cored section. However, one of the advantages of a core is that it is not just a series of spot samples, but a single continuous piece of rock, and some types of analysis seek to take advantage of this fact. An example is given by permeability analysis. Permeability is often measured by taking small plugs at specific depths, and the result is usually quoted as a permeability value at that depth in either a horizontal or vertical direction (Section 5.2.5). However, it is possible to measure permeability in a whole section of unslabbed core. If an intact length of core is chosen that is, say, 25 cm long, then the permeability value obtained will not be a spot analysis, but will be valid over that length of core. In theory, permeability could be measured over any length of core, but in practice such long sections of intact core are rarely recovered.

At the other extreme, it can be argued that a standard permeability plug analysis is not a true 'spot' sample, as it is measured on a plug perhaps 25 mm long. Within this plug, laminae may exist with greatly varying permeability. To measure at this scale, 'micropermeameters' or 'probe permeameters' have been developed (Fig. 5.4). They are mostly variants of flowmeters that measure the gas, at a set pressure, that enters the rock from a fine-tipped hollow probe pressed against it. By taking numerous measurements on an inhomogeneous core, it is possible to determine the lithological features—such as low-permeability laminae—that control the gross permeability over a greater length of core.

Suppose an exercise is carried out to calibrate an electric wireline log against a core taken over a certain interval of a well, and that the wireline log purports to give an elemental analysis of the formation. It would be possible to take spot samples of core every 25 cm, say, and to analyse them by X-ray fluorescence spectrometry (XRF) to obtain an elemental analysis. The XRF results could then be compared with the wireline log interpretation. However, this would not be a strictly valid comparison, as no wireline logs analyse a particular point (spot sample) on the borehole wall. Wireline logs record an integrated result from a certain depth interval in the well, which is dictated principally by the inherent resolution of the tool used (Section 2.8).

If the wireline log being calibrated in this instance had a resolution of 50 cm, a better comparison would be achieved by carrying out XRF analyses on a series of samples, each of which had been obtained continuously over a 50 cm section of core. The analytical results could then be compared with the electric log value obtained from the point halfway through the sampled interval (Fig. 5.5). This should give a closer comparison than with the spot samples taken every 25 cm.

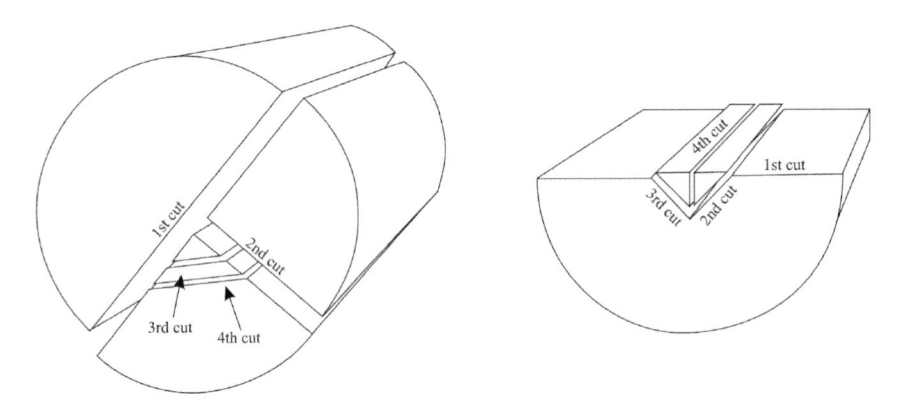

Fig. 5.6 *Two methods of using a slabbing saw to take 'continuous' core samples for analysis.*

A 'continuous' sample could be obtained by taking a number of small samples over the required interval and mixing them. For XRF analyses it would be necessary to grind the rock to a powder, and the powders from the individual small samples could be mixed to produce a homogeneous sample representing that interval. An even better method is to cut a small continuous slice from the core. For an analysis such as XRF, which would be affected by mud infiltration, the outside margins of the core should not be sampled. This can be achieved by slabbing the core twice to produce a quarter segment, and then cutting off the apex of the segment along its length (Fig. 5.6). In order to produce two overlapping sets of continuous samples, as shown in Fig. 5.5, it would be necessary to make a second cut in this position. The two prisms of rock so produced from near the centre of the core could then be split into 50-cm lengths (or other lengths as appropriate) and each length ground down and homogenized to produce a powder truly representative of that section of core.

If a rock saw with an adjustable blade is available, a more elegant method of removing a continuous slice from the centre of a core is to make two successive cuts at an angle into a half-cut of core (Fig. 5.6). The process could be speeded up by mounting two separate blades at opposing angles above a core carrier on which the half-core rests. The core slab could be passed into the path of the blades, and the sample wedge extracted in a single operation.

5.2 Analytical techniques

There are virtually no analytical techniques undertaken on core that cannot be undertaken on outcrop samples. However, some techniques are particularly appropriate for use on core samples, and others require or benefit from certain methods when used on core that do not apply to other samples. Some of the more important of these techniques are considered here.

5.2.1 Core gamma and resistivity

Measurement of gamma rays in rocks was pioneered as a method of downhole electric wireline logging (Section 2.8), but is now also often carried out on cores. Core gamma measurements are generally made by laying out the whole core, reconstructed as carefully as possible, on a conveyor belt that passes a scintillation counter measuring total gamma rays. The results emerge as a plot of relative intensity against depth.

The technique for core gamma spectrometry (Fig. 5.7) is similar to that used for conventional core gamma logging, but the scintillation detector additionally measures the energy levels, and relative proportions of the three groups of radioactive isotopes are calculated by computer and plotted against depth.

The gamma-ray reading is proportional, after making appropriate corrections, to the weight concentrations of the radioactive elements in the core. The log reading will thus respond not only to the proportions of radioactive minerals present, but also to relative thickening or thinning of the core, or missing sections. A missing section will give the same response as a clean sandstone or limestone, so it is necessary to annotate the log if features on it are caused by factors unrelated to the lithologies present.

The main purpose of running a core gamma log is usually to improve the correlation of the core with the electric wireline log suite. It is especially useful for estimating core-to-log shifts (Section 2.8) and positioning missing sections (Section 4.3.18), since thin shale beds stand out distinctly, and successions of beds result in characteristic log motifs that can simply be matched across from well to core.

Although core gamma systems generally measure radioactive intensity in arbitrary units, some have been calibrated, with allowance for different core diameters, to give results in API (American Petroleum Institute) units, which are the units in which modern wireline gamma-ray logs are presented. It is thus possible to make a direct quantitative comparison between the two.

Core gamma logs sometimes highlight features in core that are not obvious by visual inspection. It may be that a strong gamma-ray peak is noticed on a downhole wireline log, but does not appear to correlate with any lithological break at the corresponding depth in the core. The resolution of the core gamma log is a few centimetres, and if the corresponding peak can be identified in the core by measuring core gamma, the precise point on the core responsible for the anomaly can be located. A petrographic analysis of a sample from that point may recognize, for example, a concentration of detrital uranium-bearing minerals. Such a feature may form a useful subsurface stratigraphic marker horizon.

Attempts have been made to measure core resistivity, in order to correlate this with wireline resistivity logs in a fashion similar to the matching of gamma-ray logs from core to well. Unfortunately, these have met with little success. The usual technique is to place two pads or electrodes on the core, which are connected to a resistivity meter. It is found that induced fractures in the core seriously affect resistivity readings. Moreover, the resistivity measured in the borehole wall by wireline logging tools is a result of a complex interplay between the resistivity of the formation itself and that of the formation fluids, mud filtrate in the formation, mud cake on the borehole wall, and mud in the borehole.

In the core, the solid formation is present, but the various fluids and mud are either absent, or present in a totally different proportion and configuration. As resistivity in the borehole is influenced largely by the mud and formation fluids present rather than by the solid rock, it is only to be expected that core resistivity will not in general resemble wireline resistivity logs.

5.2.2 Photography

Core photography is not strictly an analytical technique, but it is commonly carried out as part of an analytical programme, and is therefore discussed here. Although a core log endeavours to communicate and record all the important information regarding a core, its purpose is not to convey exactly what a particular core looks like. Even the best logs, with full descriptions and sketches, will be unable to communicate to another geologist as full an impression of the nature of the core as would be gained by studying the core itself. Unfortunately, large sections of core are not easily portable, cannot be distributed in their entirety to more than one place at once, and inevitably deteriorate with age (either naturally or by excessive handling and sampling). If they are kept permanently at all, it is generally in some remote core store where warehousing is cheap, and visits are therefore made only in cases of special need.

Fortunately, the gap between the information which can be derived from a detailed log, and that obtainable from the original core, can partly be filled with a good set of colour core photographs. Photographs can be copied, are easily carried, and do not deteriorate significantly with age.

Almost any photograph is better than none at all. So long as the drilling mud has been washed off, photos of core laid out at the wellsite, taken with a hand-held 35-mm or digital camera, can be useful. There are few better ways for a wellsite geologist to spend any spare moments that may be available. Indeed, if the hard-won and expensive core is lost or seriously disordered in transit to the laboratory, such initiative may win an early promotion!

For long-term use, however, the best time to photograph core is once it has been cleaned and slabbed. Unlike most analyses undertaken on core, the results of which are of immediate application, photographs are of limited value while the core is still laid out in the laboratory. Any photography undertaken will therefore usually be either for illustrative purposes in a report, or for long-term records. The quality of photography required will thus depend not so much on analytical requirements as on the reporting requirements, and the perceived long-term importance of the core. Good core photography can be carried out with most popular types of camera, the most important requirement being a bright and even light source. For first-rate results, however, it is necessary to use a professional camera system mounted on a jig above a large table (or floor area) on which the core is laid out. Lighting is again a crucial element, and best results are obtained where photography is carried out in a dark room dedicated to the purpose.

Photographs of slabbed core, either half-core or resinated slabs, are generally preferable to those of whole core. Not only do structures normally show up better on the clean, smooth slabs, but it is also far easier to obtain even lighting on their flat surfaces.

Although wet surfaces sometimes show structures better than when dry, their reflectivity often causes considerable problems with the bright lights required for most photography. For routine photography it is simpler to leave the cores dry. Particular features photographed close up may benefit from being wetted, or even immersed in water or other fluid. The nature and angle of lighting can be altered. Close-up photography of geological specimens is a specialist discipline, developed largely by museum photographers, and spectacular photos of features barely visible to the naked eye can be obtained (see Bibliography). The scale of photography will depend on the size of the core, and the use to which the photographs will be put. If photography is intended to illustrate particular individual features in the core, the scale will be determined largely by the size of the feature. For routine photography of whole core sections, the optimum scale is about 1:5. A larger scale will result in an excessive number of photographs, whereas detail will be obscured on a significantly smaller scale. A useful format for recording lengths of core is to lay four to six core slabs (depending on their diameter) alongside each other so that they will lie neatly on a final print of about A4 size (Fig. 5.8). A scale, full labelling of core, and perhaps a colour chart (to adjust the colour balance of the print), should be included. An alternative format is illustrated in Fig. 5.9, where the same 1-m-long length of core is depicted in resinated slabs photographed in normal and UV light (the latter emphasizing oil staining), set against an image of the entire 360° surface of the equivalent section of whole core, taken using a specialised photographic apparatus.

5.2.3 X-radiography and tomography

Many so-called massive or 'structureless' lithologies, especially sandstones, do in fact contain sedimentary structures such as bedding or bioturbation that are not evident from a visual examination of slabbed core.

The structures may be picked out by a diffuse clayey concentration, or differential cementation, neither of which need be visually apparent. In outcrop, these structures may be highlighted during weathering, but they will be invisible in core. Apparently structureless sands can be deposited in a variety of depositional environments, and sometimes it is important to discover how a particular sand was deposited. A common method of searching for any concealed structure is to use X-rays. X-ray photographs (or X-radiographs) highlight density changes in the rock, such as occur along clayey laminae or in differentially cemented zones. Because the X-rays penetrate the rock, they 'see' structures not only on the surface of, but also within, the core. This increases the likelihood of detecting, say, a rare burrow-trace, since it will be spotted whether or not it happens to intersect the cut surface of the core. Moreover, it allows the whole of the three-dimensional structure to be seen. This often helps to determine the exact nature of a structure, and is particularly useful with bioturbation, since many ichnospecies can be fully identified only on the basis of their three-dimensional geometry. Proper identification might lead to a precise definition of the sedimentary environment in which the organism responsible lived.

Normal X-ray photography of core is best undertaken on parallel-sided slabs of constant thickness: resinated slabs (Section 3.3.3) are particularly suitable. X-rays are attenuated in proportion to the thickness of rock they pass through. Thus, if a whole-core or slabbed half-core is used, the variable attenuation of X-rays between those passing through the thick central part of the core and the thin margin will result in a photograph that is overexposed at the margins and underexposed in the middle.

If this is unavoidable, it may be necessary to build up a mosaic, comprising sections of a number of photographs taken with varying X-ray intensities or duration of exposure. It is also possible to use computer-based image processing to remove artificially varying exposure levels across the core.

In addition to X-radiography's use in revealing structures in 'massive' sands, it is also useful for investigating the nature of samples encased in some sampling medium such as wax or fibreglass (Section 5.1). A core of loose sand recovered in a fibreglass sleeve, for instance, may contain bedding that will be lost owing to disaggregation once the sleeve is removed. In this case, the difference in sample thickness between the margins and centre of the core is unavoidable.

'Real-time' X-radiography equipment is available, in which the output appears on a TV monitor instead of individual photographs. The core samples can be moved by remote control, allowing greater flexibility of use. A long length of core may be examined as it passes through on a conveyor belt, which allows rapid screening to be undertaken on a core, without needing numerous photographs. A video recording can be made if required. On reaching a zone of interest, the conveyor can be slowed or stopped, and with some machines it is possible to zoom in on a small section, and even to use an on-line image processor to enhance the image. By tilting a sample about one or more axes, the three-dimensional structure of any feature in the core becomes plain.

X-radiography is a technique used mainly in the world of medicine that has been applied to geology. A similar tool with the same pedigree is X-ray tomography, the equipment for which is best known as a 'body scanner' or 'CAT scanner'. It comprises a computer system that analyses digital X-ray photographic data built up by moving the X-ray source and detector in a wide arc around the sample (or patient). The computer builds up a three-dimensional matrix of picture elements ('pixels'), from which it can produce a two-dimensional X-ray picture of a section cut in any position or orientation through the sample. This technique is rather expensive, but has the advantage that the shape of the core has no bearing on the quality of results, so that good images can be obtained from whole-core.

Since slices of the core can be constructed in any direction, a slice perpendicular to the length of the core can produce useful information on fluid invasion into the core. X-ray tomography is also used to study the three-dimensional flow pattern of a fluid (such as oil) through a core while it is occurring. This is very valuable in the understanding of hydrocarbon flow through a reservoir.

Early X-ray tomography equipment developed for medical use worked at high X-ray energies, and was ideal for geological use. Hospitals are often happy to supplement their budgets by allowing access to their machines for an economic rent. Unfortunately, more modern machines work at lower energies that, while benefiting the patient, are less suitable for analysing rock samples.

5.2.4 Geotechnical properties

These cover a wide range of analyses commonly undertaken for site investigation purposes. Subsurface samples for these analyses would usually have been taken with an open-drive sampler such as the 100-mm-diameter (U100) sampler. Grain-size (grading) analyses are routinely undertaken by sieving; silt and clay sizes may additionally be determined by sedimentation, although these data are not routinely required. The liquid limit may be measured using the traditional Casagrande apparatus, but the cone penetrometer method is quicker and produces more consistent results. It utilizes a standard cone (Fig. 5.10), which is allowed to penetrate a sample for 5 seconds, and the amount of penetration measured. The sample is prepared by drying and passing through a 425-μm sieve; 200 g of the sample is then mixed into a paste with distilled water, and placed in a cup.

After the analysis, the moisture content is determined, and the remainder of the sample is mixed with a little more water and tested again. Once the penetration has been measured for at least four different moisture contents, these two parameters are plotted against each other to determine the moisture

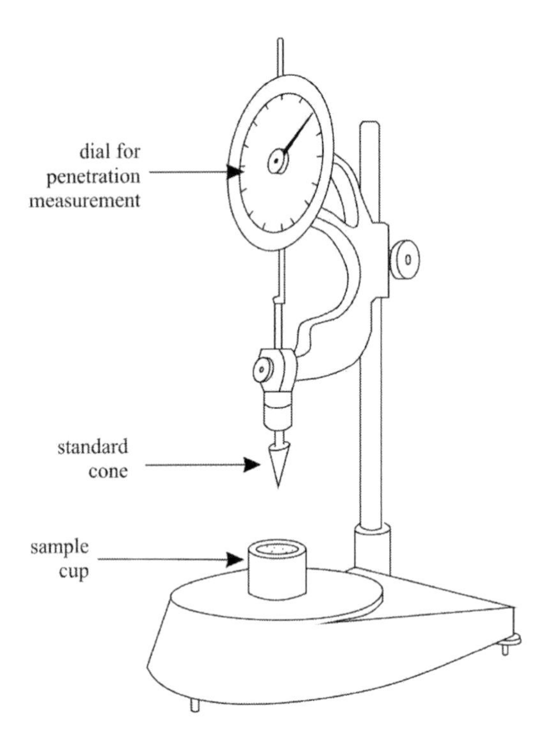

Fig. 5.10 *A cone penetrometer, used for measuring the liquid limit of unconsolidated argillaceous sediments.*

content corresponding to a penetration of 20 mm. This is known as the *liquid limit.*

The *plastic limit* is determined by preparing the sample as for the liquid limit, but then rolling out with the fingers on a glass plate to form 3 mm diameter threads, until longitudinal cracking of the threads occurs. The measured moisture content at this point is the plastic limit.

Moisture content is measured by weighing a sample before and after drying in an oven, and is expressed in per cent as the weight of water divided by the weight of dry soil. There are also quicker methods available, but oven-drying is the most reliable. Specific gravity of the sample may be measured by weighing it, then measuring the amount of displacement when it is shaken with a known volume of distilled water.

Consolidation testing measures the progressive consolidation and subsequent swell of an initially saturated but freely draining sample when subject to pressures applied by a series of weights. The sample is held in a metal ring, and therefore measures only uniaxial consolidation. A triaxial test is carried out using a cylindrical sample in a rubber sheath (Fig. 5.11). Pressure is applied longitudinally by a plunger while the sample sits in a cell of water at high pressure, which exerts a radial pressure. By measuring the stresses and deformation on the

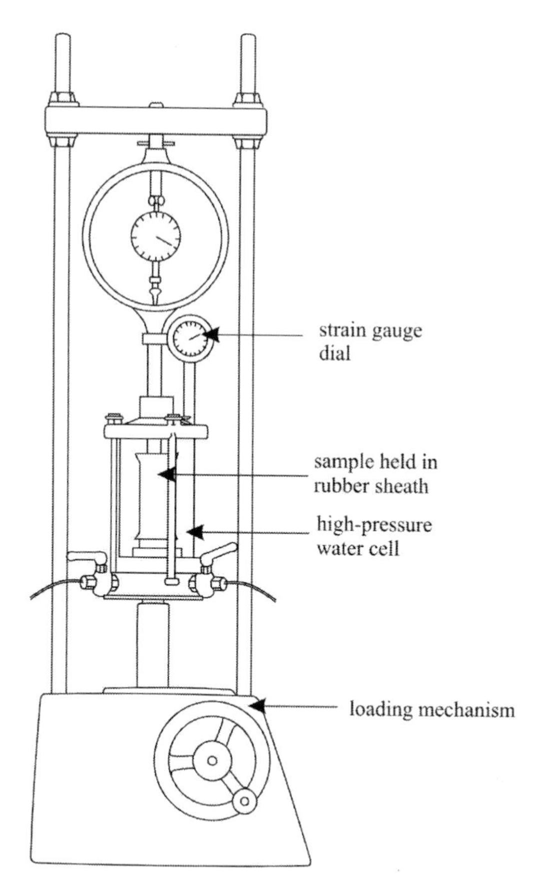

Fig. 5.11 *Apparatus for undertaking triaxial tests on core samples.*

sample, a stress–strain curve is drawn, and by carrying out tests on a number of samples, a Mohr circle can be plotted.

Permeability is sometimes measured for engineering purposes using a cylindrical sample through which water supplied from a header tank is allowed to flow. The rate of flow is used to calculate the permeability. However, since measurement of permeability from a small sample can vary considerably from actual in-situ permeability, various field tests are more commonly used for measuring permeability in geotechnical studies of near-surface deposits.

5.2.5 Conventional core analysis

'Conventional core analysis' (or basic core analysis) is a term used in the hydrocarbon industry to denote a range of routine analyses undertaken primarily to determine the capacity of a rock to store and produce (i.e. flow) oil and gas. These analyses are mostly undertaken on plugs cut at regular intervals (for example, every foot or 25 cm). Since one of the main parameters measured, permeability, varies with direction, it is common to cut two plugs at each sample point, in the horizontal and vertical directions.

Conventional core analysis is carried out largely by specialist contractors, although some of the larger oil companies have their own in-house facilities. Petroleum geologists rarely have much input to these analyses, which are strictly petrophysical rather than geological, although they can scarcely fail to notice the numerous cylindrical holes which are left in the core. The geologist will, however, be interested in the results, which effectively define which rocks are potentially of economic reservoir quality, which would form an effective seal, and which are neither. It is therefore important to have some understanding of how these results are derived.

The plug samples vary in size, but a diameter of 25 mm, and a length also of 25 mm, are typical. If the plugs contain oil, this is good news to the company concerned but bad news for accurate analyses, so the plugs need to be cleaned. This is generally carried out by Soxhlet extraction, in which an organic solvent is boiled off in a flask, and allowed to condense in a column and drip onto the plug samples (Fig. 5.12). Excess oil-stained solvent is siphoned back into the flask and reheated.

This process, which may continue for several days, allows a constant supply of clean warm solvent to pass through the plug, and is usually effective in removing virtually all traces of oil. It is important to choose a solvent (or commonly a combination of solvents) that will dissolve the oil, but which boils at a sufficiently low temperature to avoid any significant risk of damaging the mineralogical fabric (especially clay) of the rock.

Soxhlet extraction is a very common process, and is easily set up in a laboratory fume cupboard. Its main disadvantage is the long time taken. This can be critical if results will affect the drilling programme, as day rates for offshore rigs are very substantial. Alternatives include cleaning by liquid carbon dioxide in a high-pressure vessel. This method is expensive and needs specialist equipment, but can clean plugs effectively in several hours instead of days.

Once the plugs are clean and have been dried (preferably in a humidity oven), the analyses may begin. The two most important of these measure porosity and permeability. There are various ways of measuring porosity. The simplest uses a version of Archimedes' principle, involving the immersion of the plug in a liquid. The volume displaced, so long as all the pores are filled with the fluid, will be the total plug volume minus the total volume of pores in the plug. This volume is subtracted from the total plug volume, which may be calculated by measuring its external dimension, and the result is the pore volume. This value divided by the plug volume is the porosity, which is expressed as a percentage.

This method is not routinely used nowadays, since the liquid cannot be assumed to have filled all the pores without exerting external pressure or vacuum, and the technique is very time consuming. The commonest routine

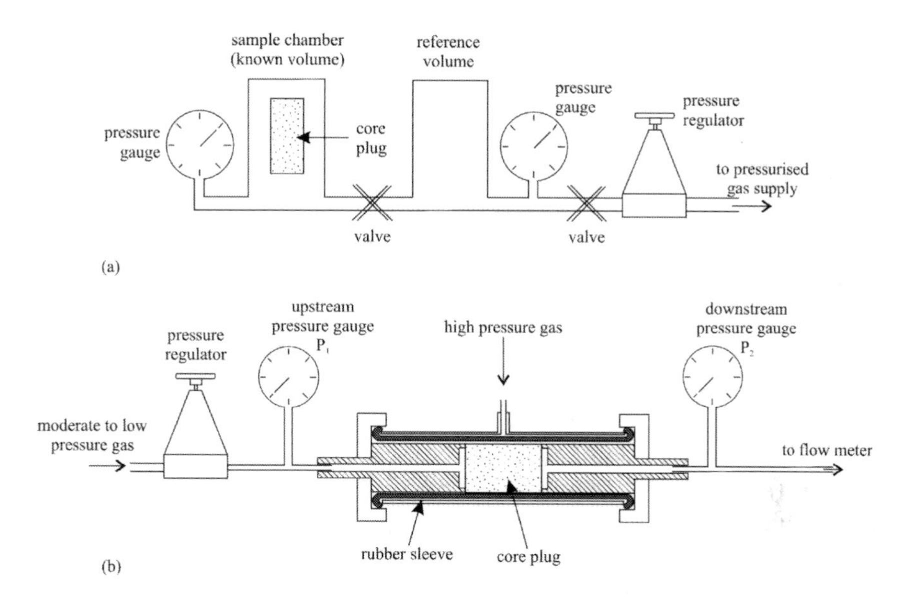

Fig. 5.13 *Schematic diagrams of (a) a gas expansion porosimeter and (b) a gas permeameter (see text for explanation).*

method uses the expansion of gas (commonly helium) from a reference chamber of known volume into a sample chamber containing the core plug (Figs. 5.13a and 5.14). According to Boyle's law, the product of gas pressure and gas volume in the apparatus remains constant:

$$P_1V_1 = P_2V_2$$

Gas is therefore admitted under pressure into the known reference volume, the pressure measured, and the valve between the reference and sample chambers opened. The gas expands into the sample chamber, and the new pressure is then measured. By Boyle's law:

$$V_2 = \frac{P_1V_1}{P_2}$$

So the volume into which the gas has expanded may be calculated. This is equal to the volume of the reference and sample chambers together, less the total plug volume, plus the pore volume. Since the plug volume is known, the total pore volume, and thus the porosity, can be calculated. The porosity is conventionally referred to by the Greek letter ϕ (phi).

Permeability is a measure of the ease with which a fluid of known viscosity will flow through a porous medium. So long as the fluid is incompressible and does not affect the medium physically, the flow rate through a sample, such as a rock plug, can be calculated by Darcy's law:

$$Q = \frac{K(P_1 - P_2)A}{\mu L}$$

where Q is the rate of flow, K is the permeability, $P_1 - P_2$ is the pressure drop across the sample, A is the cross-sectional area of the sample, L is the length of sample, and μ is the viscosity of the fluid.

Permeability is routinely measured from core by flowing a fluid of known viscosity (usually air or nitrogen) through a core plug of known dimensions across a measured pressure drop. The sides of the plug are encased in an impermeable sheath, such as rubber, which may be held against the sample by a pressure greater than the upstream pressure used to flow the air through the plug, P_1. This ensures that the flow takes place through the plug, and does not leak along its margins (Figs 5.13b, 5.15). This confining pressure, by altering the packing of the grains, may itself affect the permeability. For most realistic results it should equal the net overburden pressure that the sample experienced while underground.

By measuring the flow rate and the pressure drop across the sample, the permeability K is calculated. The unit of permeability is the darcy (D), but in

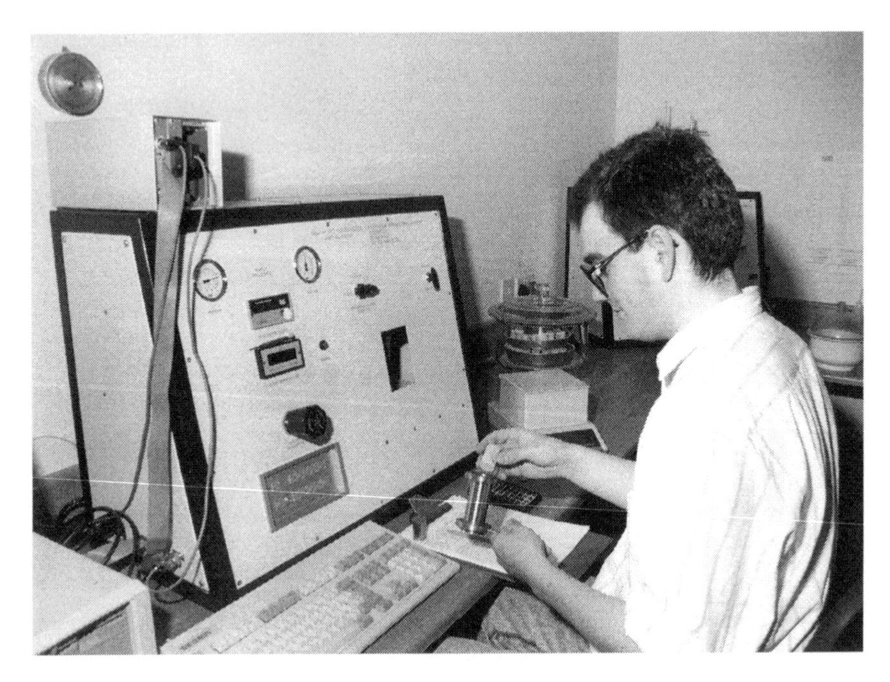

Fig. 5.14 *A gas expansion porosimeter, which uses helium to measure the porosity of gas by an application of Boyle's law. (Photo courtesy of Geochem Group.)*

most sedimentary lithologies permeability needs to be expressed in millidarcies (mD). Because the measured permeability will in fact depend slightly on the fluid used, especially if it is a gas rather than a liquid, this is usually denoted by a suffix, such as in K_{air} = 23 mD.

Another routine analysis on core is the measurement of residual fluid saturations. This measures the volumes of water, oil and gas (including air) within a sample. Although the analysis is generally undertaken on fresh core chips as soon as possible after their arrival in the laboratory, it cannot be used as a quantification of the fluid's presence in the reservoir, although it does give some indication of the relative amounts of oil present. It can be used, for example, to indicate where an oil/water contact passes through a core. In addition, by summing the three fluid phases present, a value of total pore volume is obtained. By comparing this with sample volume, an independent measure of porosity, known as 'fluid summation porosity', is obtained.

The methods adopted to measure fluid saturations are quite involved, and will not be elaborated here. They usually comprise a combination of mercury injection and retorting.

Conventional core analysis results often include a figure for 'grain density', which is a measure of the average density of the mineral content of the plug (that is, not including porosity). Since the total grain volume has been measured during porosity determination, the grain density is easily calculated by weighing the plugs. The value can be used to calibrate electric logs, some of which measure, or are dependent on, formation density. It is also a useful indicator of anomalous plugs. If a number of limestone plugs with a density of around 2.71 g cm^{-3} contains one with a value of 3.30 g cm^{-3} it demands a closer look. It may, for example, contain a pyrite nodule, which would also affect other analytical results on that plug.

Various other analyses are sometimes grouped under the heading of 'conventional core analyses', but those mentioned above are the most important from a geological viewpoint. Techniques described elsewhere, such as core gamma measurement and core photography, may also be included. Indeed, the term is sometimes used to describe all of the operations routinely undertaken on oil company core once it arrives in the laboratory, including slabbing and resination.

However, it applies primarily to those analyses undertaken to measure the hydrocarbon storage and production characteristics of a reservoir rock.

5.2.6 Special core analysis

In addition to the 'conventional' analyses undertaken on oil company cores, there are numerous techniques undertaken in special core analysis laboratories. These are widely known as 'SCAL' analyses. In addition to core chips and plugs, preserved samples of whole core (or 'full diameter core') are widely used in special core analysis: hence these samples are commonly designated

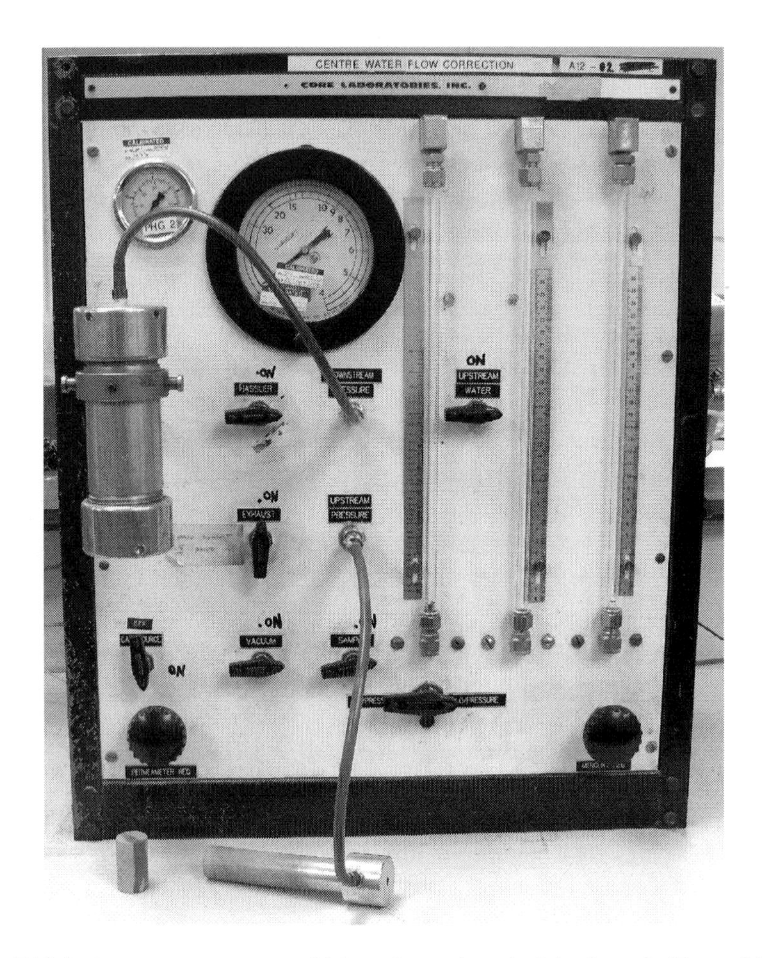

Fig. 5.15 *A nitrogen permeameter, which works on the principle shown in Figure 5.13b, and measures permeability according to Darcy's law. The metal cylinder on the left holds the core plug. (Photo courtesy of Core Laboratories Ltd.)*

'SCAL samples'. A wide range of analyses may be undertaken. In addition to further analyses of porosity, permeability, etc., including 'relative permeabilities' when two or more fluid phases are present, there are studies of how the formation itself is affected, including the effects of injecting various fluids that might improve (or indeed damage) reservoir quality or hydrocarbon recovery.

Certain common geological analyses, such as optical petrography, may be included under 'special core analysis', because to the petrophysicist or petroleum engineer who usually commissions SCAL studies these are specialized techniques. The term also indicates that these analyses are not routine (or cheap), but are undertaken to address specific problems that may have been encountered in a particular reservoir.

The petroleum geologist will usually have little direct involvement with SCAL work. The main impact of these studies on the geologist often arises from the fact that significant sections of whole core need to be preserved in wax (or by some other method) at the wellsite, and are therefore unavailable for immediate geological appraisal.

5.2.7 Other analyses

There is a considerable number of analyses undertaken on cores that are grouped here as 'other', not because they are less important than those noted above, but because they do not apply only to cores, and are not carried out in a significantly different manner on core than on outcrop material. These techniques are widely discussed in the geological and allied literature. Petrographic analyses, utilizing thin sections, the scanning electron microscope (SEM), and X-ray diffraction (XRD) are examples of techniques that are regularly used by the geologist working on core, but which do not differ from the same techniques used on outcrop samples. The same goes for biostratigraphic studies, such as those utilizing palynology and micropalaeontology. Although fresh core is an ideal source for biostratigraphic material, the methods used in sample preparation and analysis are sufficiently well documented elsewhere.

6 Interpretation and preparation of final logs

6.1 Logging and interpretation

The description of a particular man may mention a scar on his left cheek. If, however, it was therefore reasoned that the same man had been badly cut on his left cheek at some time in the past, this would be more than a pure description, and would be classed as interpretation, even though it could be directly inferred from the description. It is important when logging core to distinguish, as far as possible, between observation and interpretation. For example, for the sake of clarity and brevity, it is justifiable to describe a particular structure on a log as, for example, chicken-wire structure in an anhydrite sequence rather than to describe it in detail. The inference 'sabkha environment', however, is more than a description, and should be reserved for a quite separate 'interpretation' column. It is then possible for future readers of the log to distinguish precisely what was logged, and to follow the logger's reasoning that led to a particular interpretation. In fact, in completing an interpretation, the geologist is no longer acting as a logger, but has taken on the role of, say, a geotechnical engineer or a sedimentologist. Interpretations are rarely as clear-cut or unequivocal as that chicken-wire anhydrite is formed in a sabkha environment (and even this, no doubt, has its exceptions). Descriptions are concerned with those features that are physically present in the core, whereas interpretations are inferences as to how they got there, or how they would respond in certain circumstances, such as trying to produce oil from them, or excavating a motorway cutting beside them. An effort should be made to keep the two separate.

Core logs will differ considerably in style or content from one another, but so far as possible they should record observations, whereas an interpretation, while perhaps appearing side by side with the log, records deductions.

The above should not be taken to mean that the logger must steer clear of interpretation, but just that the logging and interpretation should as far as possible be separated, both in the logger's mind and on paper. Indeed, no one is better placed to make an interpretation than the logger while in the presence of the core. Although interpretation may be refined by subsequent laboratory analysis, the logger should record, with reasoning, any initial interpretations while logging is in progress, together with any thoughts or hunches that may seem relevant. It is astonishing how one's mental view of a core changes after it has been logged and put away. All sorts of features that were obvious at the time are forgotten. One's memory becomes even more restricted once a particular interpretation has been adopted. It is very easy to forget all those features in the core that do not fully accord with the current model, so that after a while one's mental image of a core becomes quite different from its actual appearance. For this reason, it is particularly important to make a written interpretation of a core while it is being logged, but paying special attention to any difficulties with that interpretation, and any plausible alternatives. This also emphasizes the value of working subsequently not only with a good log, but also with a good set of colour core photographs to serve as a constant reminder of the actual appearance of a core.

It has earlier been emphasised that logging should be carried out methodically, and the same applies to interpreting a log. The advantages of core over outcrop material have been noted above. Another advantage is that core is ideal for working methodically. A geologist mapping the countryside has all manner of outcrops of different shapes and sizes, and never knows when or where more outcrops will be uncovered. Sampling strategy in this case tends to be virtually random, or dictated largely by the position of the base camp or hotel. Even sampling on a defined grid is usually greatly influenced by the vagaries of exposure and weathering. Worse still, the exposures that do exist are usually composed of the most durable lithologies, so are not even representative. In contrast, a core (assuming reasonable recovery) is a representative sample through a given section. It is of defined size, the various lithologies follow in a strict sequence, and samples can be defined (usually uniquely) by a single number (namely depth). Sometimes two or more cores are available, but they will both be in this useful form, and it may even be possible for the geologist to decide where to site a further well. In other words, core is close to ideal for working systematically and 'scientifically'. Many geologists, trained on the outcrop, do not derive the full benefit from this 'neat packaging' of core material.

Although interpretation has been considered so far as though it were a single-step process, it can be split into three quite distinct phases, which must be followed sequentially. Much unscientific work and confusion have arisen on occasions when these steps have not been followed through systematically and, for example, where the geologist has moved straight through to the final 'definitive solution'. The first of these interpretative steps involves the simple recognition and naming of a particular lithology or structure.

The second phase is the interpretation of that feature—how it formed, and its consequences with regard to the work in hand. This second stage deals with interpretation of individual features that are actually present in the core, and it therefore may be referred to as interpretation by internal evidence. The third phase uses the results of the first and second stages to build a geological model that extends beyond the limits of the core itself, and may include aspects that are not represented in the core at all. For example, a core of sediments, recognized during the first phase as passing down into a zone of induration and hornfels texture, may be interpreted in the second stage as having been thermally metamorphosed. The third stage of interpretation may invoke the presence of an igneous intrusion close to the lower part of the core. This final interpretation gives rise to a geological model containing features (the intrusion) not present in the core at all, and it may thus be referred to as external inference.

All core logging necessarily includes the first stage of interpretation, and may or may not lead to further stages. The following sections discuss these three levels of interpretation in turn.

6.2 Recognizing and naming features

It has been emphasised in Section 4.2.1 that even the most basic logging can hardly be undertaken without some degree of interpretation. Even marking the depths on the core will often require some interpretation to decide the positions of missing core sections, and the position and orientation of core pieces that were misplaced while extracting the core from the barrel. And geological descriptions are largely interpretative, as a glance through the list of symbols and abbreviations in Appendix 1 demonstrates.

If, for example, the logger recognizes desiccation cracks, they should be recorded on the log as such. After all, another geologist looking at the log and reading 'mudstone bed containing nearly vertical upward-widening wedges composed of sandstone identical to and continuous with that in immediately overlying bed' would need to think hard and may well fail to recognize from this description a structure that any experienced geologist would recognize as formed by desiccation cracks. Yet the term is quite clearly not simply a description, but an interpretation as to the mode of origin of the structure. Core logging, and geological description generally, is not just a descriptive art, but an interpretative science.

Yet although the presence of desiccation cracks, or open fractures, or contemporaneous weathering horizons within lava sequences, have clear interpretational implications, it is not the job of the logger, *acting as the logger*, to draw these implications. Although geological descriptions are largely interpretations, this interpretative aspect should be taken no further than is necessary for clarity. It is vital that the basic log is as close as possible to being an objective description of the core, with an emphasis on creating a faithful record. Interpretations are always more prone to error than descriptions. It may be necessary to log 'desiccation cracks', but it is not useful at this stage to add 'resulting from a prolonged dry spell'. Perhaps the cracks formed beneath the shelter of a tree during a downpour!

Unfortunately, some of the most profound differences in final interpretation arise from uncertainties at this earliest stage. One geologist may log a certain structure as a rootlet trace, implying that the overlying coal accumulated *in situ* in a subaerial setting. Another geologist may interpret the same structure as a vertical burrow indicative of a marine shelf setting. Once the log is drawn up, the user is likely to take it at face value, unless any uncertainties present have been made clear. It is therefore imperative that a log is not simply a 'best guess', but that the logger, while perhaps drafting the preferred interpretation on the graphic lithological log, also makes a note of any such uncertainties or alternatives. Otherwise, the user who interprets a sequence as a deep marine fan, but is concerned by a core log recording halite pseudomorphs within the sequence, will have no indication as to whether his interpretation or the log is more likely to be in error.

6.3 Internal evidence: interpreting the core

So the first stage of interpretation is limited to the minimum required to produce a core log that is clear to another geologist. So far, there has been no scope for a designated 'interpretation' column on the log. The second stage is to produce a full interpretation of the core as a whole, which should seek to explain all the features described in the log, so far as is appropriate to the project in hand. This explanation is the crux of the second stage of interpretation. The explanation may be in terms of the depositional (or igneous, metamorphic, etc.) environment responsible for the 'observed' lithology, or of the manner in which fractures have formed, or of the chemistry and sequence of fluids responsible for mineralization.

It is not intended here to assess all of the geological techniques used in interpreting the core. They are essentially the same as those used in outcrop studies, so they are part of every geologist's training, and are widely covered in the geological literature. As has been discussed elsewhere, core differs from outcrop, and there are various advantages and disadvantages of one over the other. However, the principles of interpretation are the same, requiring only a different way of thinking, which comes with experience rather than with retraining.

An example of this different mode of thinking is given by methods of interpreting depositional environments. It is very common in the sedimentological literature for depositional environments to be characterized by a certain vertical facies sequence, sometimes called a 'facies model' (see for example Anderton, 1985; Reading, 1996; Posamentier and Walker, 2006) These are usually summarized in the form of a sedimentological log. Of course, a core log through a sedimentary sequence may be drafted in precisely the same format. When endeavouring to interpret a sedimentary environment from core, it is tempting, therefore, to scan through a textbook of facies models until one is found that closely matches the core log. This is a rather dangerous practice, however, as the two types of log have been constructed according to two quite separate approaches: one derived from outcrop and the other from core. The facies model is usually constructed by studying one or more surface outcrops, measuring a number of representative vertical sections, and then 'distilling' these into a model section that is considered to be an idealized representation of the sedimentary sequence deposited by that particular depositional system. It is probable that no section through the outcrop actually displays that particular sequence, and many sections may be totally different. A facies model may, for example, display a broad upward coarsening, although numerous actual vertical sections cut through the sequence it represents may be upward fining.

The core log, in contrast, depicts an actual, virtually random, section through a sedimentary sequence. The chances of it being typical of the sequence from which it was cut are small, unless the sequence is one of particular lateral uniformity. Indeed, given a textbook containing, say, 20 facies models, a random cored section through a sedimentary sequence is probably less likely to resemble the model section most appropriate to the environment in which it was deposited than it is to resemble one of the other 19. Facies models are certainly useful in that they summarize the range of lithologies and structures that may be associated with a particular environment. They must not, however, be considered in general as models for sequences to be found in core.

As noted above (Section 6.1), the geologist logging core is in an excellent position to make interpretations, but interpretations should not form part of the basic log. It is an excellent idea, however, to add a column on the far right of the preliminary logging form headed 'Interpretation', on which the logger may record ideas as they arise.

6.4 External inference: geological modelling

Up to this stage, an effort has been made to describe and explain the core. Paradoxically, the core is the one part of the formation that does not need to be described and explained, since it is no longer underground, and will play no future part as an oil reservoir, load-bearing medium, or whatever property of the rock is being assessed. Of course, the core is not cut for its own sake, but as a representative sample of the formation penetrated. Volumetrically, the sample may appear wholly inadequate. If an oil reservoir is assessed by cutting 10-cm-diameter cores on a 1-km grid pattern, for example, the resultant core will form less than a hundred-millionth of the reservoir volume. Even this assumes that the cores all penetrate the whole reservoir, and recovery is 100%.

Fortunately, geological units mostly have some degree of lateral continuity, and if a core penetrates a certain succession it is likely that a similar succession exists laterally for some distance around the borehole. Of course, this is not always the case if the well happens to penetrate, say, a volcanic plug or a fault zone, but it is to be hoped that the geologist will recognize these features and treat the core data accordingly.

This is the third stage of interpretation: proceeding from the explanation of the features recognized within the core to inferring the nature of the surrounding formation. If a study involves core from more than one well, it is at this point that the interpreted data from the individual wells are correlated and integrated. The assessment moves on from the features observed in the core to those inferred beyond the well-bore. This is done by using the interpretation of the core or cores to build up a geological model that is capable of predicting the nature of the formation at any point within the area of investigation. The geological model may well predict features that are not observed in any of the cores. If, for example, one well contains a core that is interpreted as having been deposited by beach sands, and a second well some distance away has a core at the same stratigraphic level that contains marine shelf muds, it may be inferred that a marine shoreface succession lies between the two. There may of course be additional data, such as seismic reflection lines, that will assist with this modelling.

Because this interpretation is not just an interpretation of the core itself, it would not usually appear on the core log. Indeed, much of this type of interpretation is best expressed in map form. In many cases this interpretation, together with the reasoning behind it, will be outlined in the text of a report on the study undertaken. Appended to the report will be copies of the final core log, together with maps and other diagrams as appropriate for illustrating the conclusions.

6.5 Core stratigraphy

Another aspect of core interpretation is the recognition and definition of stratigraphy. The scale of stratigraphy will depend on the nature of the work being done. A shallow borehole sunk for site investigation purposes may penetrate 0.5 m of topsoil, 1.5 m of windblown sand and 2 m of till. If a similar succession is encountered in several boreholes on the site, this would be regarded as the stratigraphy of that site, and would form a framework for site investigation. On a larger scale, a deep hydrocarbon well may penetrate 1000 m of limestone and 500 m of shale and terminate in 250 m of granite. Comparison with units of similar scale in adjacent wells would lead to a broad stratigraphic subdivision of that particular basin. The subdivision could be refined by recognition of marker beds within the larger lithological units, or by the use of biostratigraphic zonation.

The primary local stratigraphic unit is the formation, which should possess some lithological homogeneity or distinctive lithological features (petrological, mineralogical, geochemical or palaeontological) that allow it to be mapped on the surface or traced in the subsurface. Formations are usually defined with reference to a type section in outcrop, but in the absence of surface exposure a type well may be used. A type well is chosen and described to demonstrate the distinguishing features of the various units and their boundaries.

With one-off studies it is unlikely to be worthwhile to define the stratigraphy formally, but if several studies are likely to be carried out in the same area, and especially if they will involve several different groups of geologists, early definition of stratigraphy can solve many communication headaches.

6.6 Why a final log?

The final log is essentially a document for presentation purposes, as opposed to the preliminary log, which is that produced by the logger while working on the core. In many cases the preliminary log will be the final log, either because a log is not specifically needed for presentation purposes, or because the initial log is sufficient, perhaps after professional drafting, for this purpose.

In other cases, though, the format of the preliminary log is unsuitable for presentation. The preliminary log may be too detailed, or may need to be amended or added to in the light of analytical data. It may be at an inappropriate scale, or may be best presented as part of a correlation diagram with other logs.

Final logs will not usually supersede preliminary logs, but are produced for a different purpose. The preliminary log is intended to be a comprehensive record of a core (at least with regard to the requirements of any particular core study), whereas the final log may highlight aspects that are considered to be particularly significant. The preliminary log will often contain an interpretation column in which the logger has recorded hunches and speculation, much of which may subsequently be judged to be irrelevant.

The significance of features originally recorded by a logger may, however, not become apparent for years, and will not be transferred to the final log. Thus, even though a final log may be produced and neatly drafted, any hand-drawn preliminary log containing logger's comments that are for the time being considered unimportant should be carefully filed for future reference. As a long-term record of a core, the original log, perhaps creased and coffee-stained, can sometimes form a much more valuable record than an edited and polished log prepared for client or management consumption.

We observed earlier (Section 4.1) that this is one of the disadvantages of entering data straight onto a computer during logging, so as to produce a digital final log that does not require any redrafting. There is much less likelihood in this case of the geologist recording potentially useful ad hoc comments. Of course, such systems rarely record fine detail, so they are really suitable only for quite simple and routine logging. However, where this is the requirement, they do have the advantage that a clean, final log can be printed out immediately on completion of the job.

6.7 Format of the final log

The final log can be presented in a wide range of possible formats, some of which bear little resemblance to the preliminary log. Some points to be considered in drawing up a final log are discussed below.

6.7.1 Scale

The choice of scale for the preliminary log was discussed in Section 4.2.2. Redrawing a log in detail from one scale to another can be very time consuming, so it is as well to ensure that, wherever possible, preliminary and final logs are at the same scale. This is not always possible: a client may have a change of mind halfway through a job, or a summary final log may be required at a small scale (say 1:200), although original logging and subsequent interpretation may have required a larger scale (say 1:20). There is rarely any value in redrawing a preliminary log at a larger scale, as it will tend to appear very sparse in detail. This will usually be worthwhile only if it is to match the scale of another data set that is too detailed to be matched to the preliminary log scale.

Small changes in scale (up to perhaps a factor of two) can be attained quite simply, as it is usually possible simply to expand or contract every detail of the log. It may be possible simply to alter the scale of a digital image or photocopy. This is unlikely to be satisfactory with greater reductions in scale, since the resulting log will be too condensed for the detail to be understood. In this instance it will be necessary for the geologist to draw a summary log, which will omit the less significant detail. This may involve the merging of beds, omission of detailed structure, and 'averaging' of grain size variations. Achieving this while still representing accurately the true nature of the core can require careful thought (Fig. 6.1).

6.7.2 Correlation diagrams

In addition to the straightforward core log, one of the most common methods of presenting core data is as part of a correlation diagram (Fig. 6.2). This comprises a horizontal section through the formation, passing through two or more bore holes. Each well on the diagram is represented by a graphic well log, and the areas between the wells usually contain a geological interpretation, which may

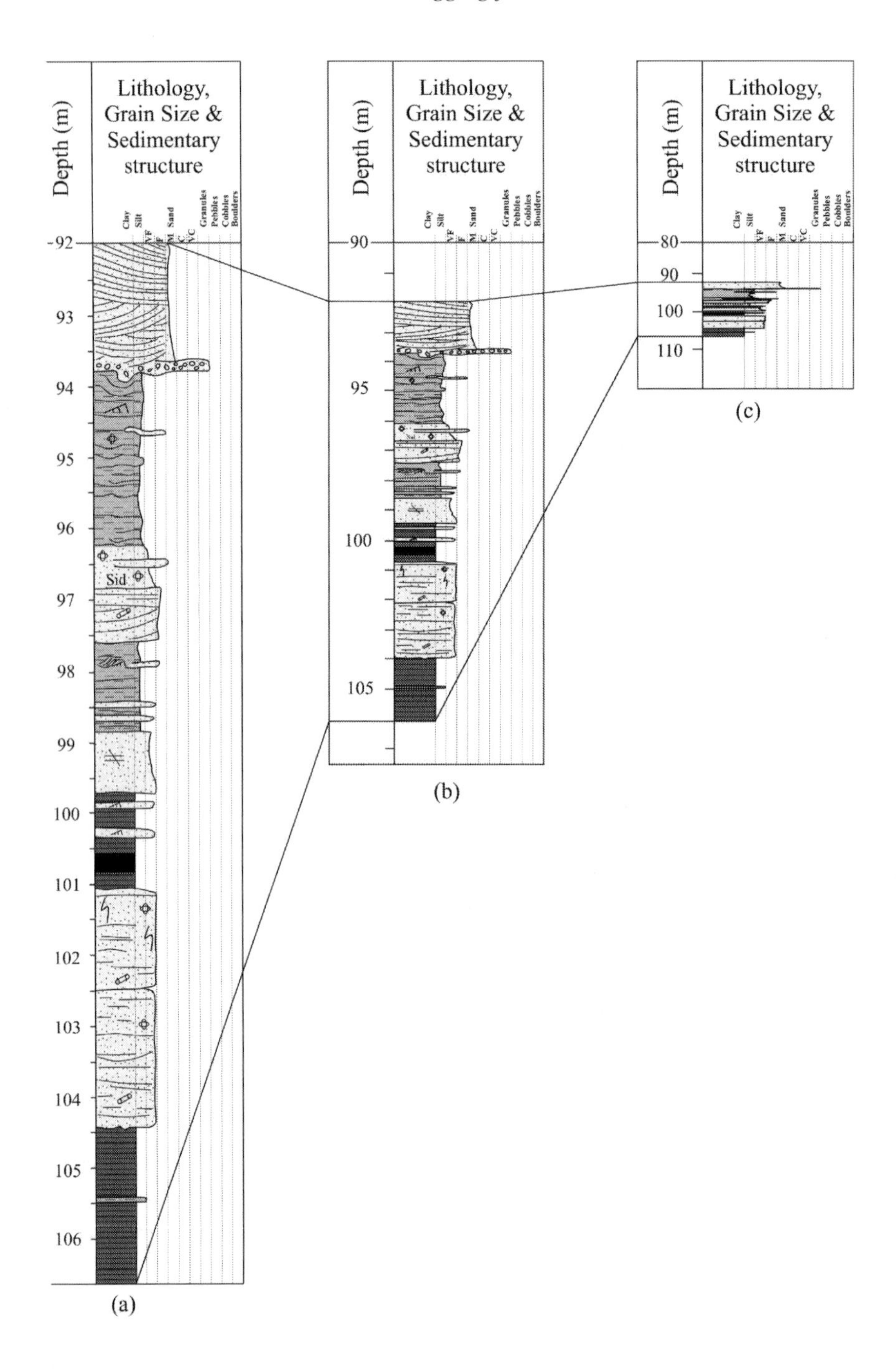

Fig. 6.1 *The same section of core depicted at three different scales, illustrating how data need to be condensed and generalized at smaller scales. The scale of log (a) is 20 times that of log (c).*

be based on the well data alone, or with additional surface mapping or geophysical data.

Core logs used on correlation diagrams are usually presented in a simple format, such as a single column, so that the message of the diagram as a whole is not obscured by too much fine detail. For shallow site surveys with a close spacing of wells it may be possible to produce a diagram with no vertical exaggeration. This is ideal, since it correctly reproduces angular relationships in the formation. In practice, however, it is usually necessary to introduce a considerable vertical exaggeration.

Correlation diagrams in the form of two-dimensional horizontal sections can be produced realistically only if the wells fall roughly in line. This will often not be the case, as when a grid pattern of well locations has been used. In this event a fence diagram is more appropriate, or just a map or ground plan with a simple core log adjacent to the position of each of the well locations.

6.7.3 Analytical data

A final core log is often accompanied by analytical results obtained from the core. Indeed, the core lithology log may be just one of numerous collections of data derived from the core, and may form just a small section of an expanded core data sheet. Methods of plotting data against a core log are summarized in Section 4.2.4. Large core data summary sheets certainly form a handy repository for a substantial quantity of information relating to the core. However, they can be unwieldy, and important items of data can become buried amongst a crowd of more obscure analytical results that would best be consigned to the appendix of a report. Ultimately the choice of how to display data is largely a question of personal preference.

As described in Sections 2.8 and 4.3.2, there are sometimes several methods used to measure depth in a deep well. Core depths may differ from electric wireline log depths, and both may differ from true vertical depth. In addition, different datum points may be used, such as ground level, sea level or kelly bushing. In constructing a summary log that incorporates various data sets it is important to make allowances for these different depth measurements.

Electric wireline log depths in particular rarely match core depths in deep wells, and it is usually simplest to have two separate depth scales, clearly marked, on the summary log, so that there is no confusion as to the relationship between the different systems (Fig. 4.10). If electric wireline logs are available from a well, and unless particular reference is being made to a core, it is wireline log depths that are usually quoted.

Another item that is essential on some types of log is stratigraphic information (Section 6.5). This may be derived simply from prior knowledge of the formation drilled, by lithological matching (lithostratigraphy), or from biostratigraphic dating (biostratigraphy). In some cases absolute dating may be carried out on core material using, for example, radiometric or palaeomagnetic methods. The stratigraphy is commonly documented in a column forming the far left-hand side of the log. Several columns may be required in the case of a longer cored section to include several levels of a stratigraphic hierarchy. Where the stratigraphy is based on general regional knowledge, this may be explained in an appropriate section of the log (such as the 'comments' or 'interpretation' column). Where it is based on specific analytical results, it may be useful to include these on the final log. Detailed biostratigraphic correlations, especially those using micropalaeontology or palynology, can be very complex. These correlations can be very sensitive to lithologies, so the data are commonly plotted alongside a lithological log to form an illustrative data compilation in their own right (Fig. 6.3).

A great advantage of producing logs using modern digital methods is that it has become very simple, where core photographs are available, to insert a column containing the photos alongside the core log. The two usually complement each other very well, and provide a much clearer visual representation of the nature of the core than either the log or the photos would do in isolation. Ideally, if the log is prepared at quite a large scale, the photos can be reproduced at true scale. On smaller-scale logs core photos at true scale become so narrow that not much can be seen of them. In this situation the photos can be 'stretched' horizontally to widen the column: dip angles will be reduced below true dips, and this should be noted on the log, but the method can be effective in improving the visibility of features in the core. Of course, at very small scales the photos become meaningless even with horizontal exaggeration, but it is often surprising how informative a series of core photos can be when presented against the log of a long length of continuous core, even at scales as small as, say, 1:200. Only the very largest individual structures are visible, such as large-scale cross-bedding, but if the core comprises several lithologies with distinct colours, oil staining above a certain oil/water contact or other similar visual features, their depiction can significantly enhance the value of the log.

6.7.4 Legend and header

In addition to the columns of data, final logs should usually incorporate a legend and header. In the simplest logs, though, the legend may be self-explanatory and can be dispensed with. Obviously, if several logs are bound into a report or publication, there is no need to repeat the legend on each log. Similarly, some institutions have their own in-house standard legend, which does not need to be repeated on all logs intended only for internal distribution.

The legend should comprise an explanation of all symbols and abbreviations used on the log, although any abbreviations that are common knowledge or are self-evident may be omitted. As Appendix 1 shows, a legend could be very lengthy, although few logs describe a sufficient range of lithologies or are

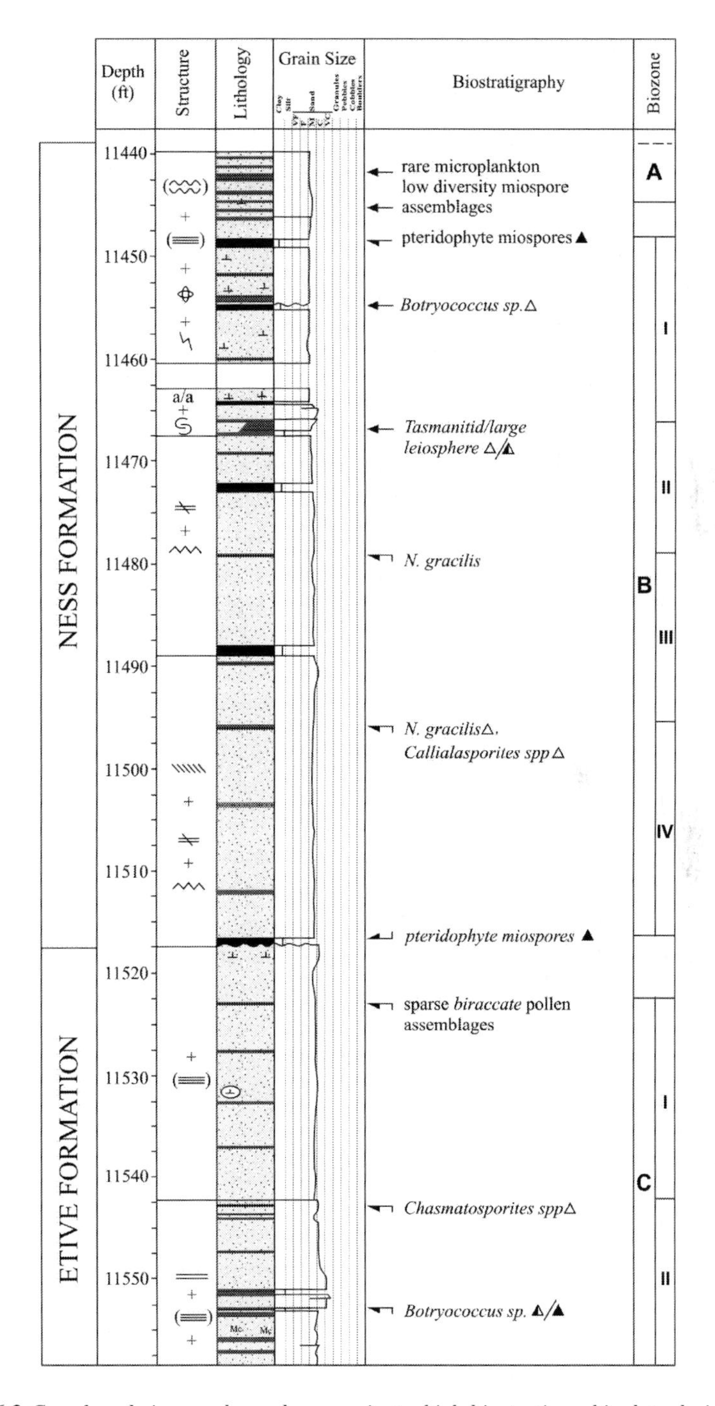

Fig. 6.3 *Core logs being used as a base against which biostratigraphic data derived from core samples have been plotted. This can be very useful, especially as the biostratigraphic information may have been affected by the sedimentary facies.*

sufficiently detailed to need to be as extensive as this. Final logs are, however, often drafted on standard bases. A legend included on such a base designed to cover all eventualities would be prohibitively long, whereas to draw a separate detailed legend on each log would be very time consuming. A reasonable compromise is to have a pre-drafted legend covering all of the most common symbols, but with a few spaces enabling any less common symbols used on any particular log to be inserted.

The header is basically the title of the core log, together with other information about the well (Fig. 4.8). In addition to the name and location of the well, this may include the name of the institution, company, contractor or client responsible for the well, and its interpretation, together with the name of the author of the log. Any significant information about the well, such as the datum from which depths are measured and the drilling or coring method used, should be included, together with the scale of the log, the date the logging was undertaken, and any other relevant remarks. If details such as the type of drilling rig, core barrel and bit, drilling progress and contractor are not otherwise available on file, they can be recorded (if known) on the core log.

6.8 Final drafting

A detailed core log can be a draughtsman's nightmare, with so much to pack into so little space, and such a complexity of symbols, freehand drawing and text. Because of this, some of the neatest and clearest logs are completely hand-drawn and handwritten. But often, owing to company policy or publication requirements, a log must be professionally drafted. Nowadays, of course, this usually means computer-drafted. However, whether drafted manually or digitally, a final log drawn up by a professional draughtsman will look as neat as the best hand-made variety only if it has been carefully thought out. Furthermore, it is likely to be precise only if the drafting is conducted with some knowledge of the geological principles involved.

A variety of different text sizes and styles will need to be used. The available space on a log for comments may be small, requiring a small letter size, whereas a short description relating to a long cored section of uniform lithology will need to be written large and boldly to draw attention to itself. The content of a graphic lithology column is rarely composed of straight or even regular lines. A draughtsman's job is often to 'neaten up' a hand-sketched original, but many a realistic grain-size curve, for example, has been the subject of a tidying-up exercise by a draughtsman imposing straight lines and segments of circles on the real world.

Similarly, standard patterns taken from a 'menu' can be used for the representation of simple lithologies, but where a complex lithology occurs, or there is fine interbedding, it is much simpler to revert to freehand drawing or its digital equivalent. Standard symbols can also be cut and pasted from a menu, and can be resized digitally, but here again a freehand representation of structures and other features of the core is often much more useful, and the geologist must ensure that the final log does not become over-stylized. A draughtsman's training will often lead him or her to resort to the most automated, least time-consuming methods, and sometimes they must be encouraged instead to draw on their creative and artistic talents.

If care of this type of is taken, a professionally drafted log can be produced, and still retain all the detail and subtleties of expression intended by the geologist.

7 Core preservation and storage

Some cores are cut, described, and disposed of. Others are cut, described, and consigned to storage in perpetuity. The former course of action may turn out to be expensive if data derived from the core are lost or found to be incorrect, or if further information is required from the same site at a later date. The latter course is guaranteed to be expensive once commercial rates of transportation, curating, storage and maintenance are taken into account. Although from an academic viewpoint it is desirable to preserve all core (and other well records) indefinitely, there are strong commercial arguments for not doing so unless future work on the core is anticipated.

Nevertheless, it is worth pointing out that government bodies (such as the BGS in Great Britain and the USGS in the USA), in addition to universities, are often very grateful to accept cores, especially from deep wells. Good-quality core is generally not easy to obtain, even for teaching purposes, and various institutions will be happy to use core that would otherwise go to waste.

Having decided to invest in the long-term storage of core, it is only sensible to ensure that the core will remain in as good a condition as possible, so that it will remain a usable asset. In practice, this depends not only on the core remaining physically intact, but also on its being properly documented so that core from a particular well and depth can be readily traced. This chapter deals primarily, however, with the physical preservation of core.

The term 'core preservation' is used in two ways. It is used first to denote the various methods used at wellsite, such as sealing in wax, to keep the core as close as possible to the condition, especially with regard to contained fluids, in which it existed downhole. If the core is not specially preserved in this way it quickly equilibrates to surface conditions, principally by drying out, but is otherwise reasonably representative of the lithology downhole. In the context used in this chapter, however, preservation refers to the methods of inhibiting the various processes (natural and artificial) that will lead gradually to the deterioration of the core from this latter condition. It needs to be remembered that, unlike outcrop samples, core samples have often come from depths where physical and chemical conditions allow minerals to survive and grow that are unstable in the surface environment of temperate climates. Positive measures therefore need to be taken if deterioration is to be limited.

This chapter is included primarily in the hope that the geologist may have some influence as to the manner in which cores are stored. This is not always so, however, in which case the details given here may need to be used by the geologist to explain how the damage and deterioration apparent on the archived core being logged has occurred.

7.1 The core store

Core can be, and often is, preserved satisfactorily in all manner of unfavourable conditions. Some types of rock are, after all, extremely durable, and form high mountain ranges that stand as symbols to unchangeableness ('Rock of Ages'). However, even apparently durable rocks undergo various subtle changes when exposed to the elements, and other rocks deteriorate rapidly. If a choice is available, it is worth ensuring the best possible conditions for core storage.

Most large core stores are essentially warehouses with metal racking (Fig. 7.1). If large quantities of core are to be moved regularly, this is best achieved with a forklift truck. This will, however, require wide passages between the

Fig. 7.1 *Core storage in a large modern warehouse. The shallow trays are used to hold resinated slabs. (Photo courtesy of BP Exploration.)*

shelves to gain access, unless the racking is of the mobile sort and can be moved laterally on rollers to maximize usable space.

Most core stores are less sophisticated than this, but so long as there is space to store the core boxes horizontally, and they can be stored and recovered without difficulty or danger (either to the cores or to the personnel—rocks are heavy) the precise layout of the core store is not critical.

Environmental conditions can significantly affect the chemical stability of rocks. Unfortunately, the only such condition that can be controlled satisfactorily in most core stores–temperature–is of relatively little importance. If possible, temperature should remain constant, and this is more important than achieving any particular temperature level, although the rate of chemical reactions that do occur in rocks will tend to increase with rising temperature. An average of 15–20°C is about right, within a range of about 5–25°C. Much more important than temperature, however, is relative humidity, which unfortunately is seldom controlled in core stores. High relative humidity is responsible for rapid oxidation of pyrite and other common sulphides. The oxidation leads to the disintegration of the mineral itself and damage to the surrounding rock. This reaction occurs very rapidly at a relative humidity of 60%, but sulphides can remain in good condition for several years or more once the relative humidity is reduced to 50%. Because of the widespread occurrence of pyrite in subsurface rocks, its deterioration is a very common sign of inadequate core storage conditions.

Some minerals are hygroscopic, the most important being the smectite clays (such as montmorillonite), which occur in many shales. Fluctuating humidities result in their successive swelling and shrinking, causing cracking, distortion and disintegration.

Cores drilled with saline mud can develop a surface bloom of salt crystals due to constant hydration and dehydration. Calcareous cores can develop a similar-looking problem, with numerous thread-like crystals on their surface (this is known as 'Byne's disease'). Fortunately, both these surface efflorescences can usually be removed by washing the surface, but this is at best inconvenient and may disturb more delicate core materials.

Some water-soluble minerals, especially halite, tend to deliquesce or absorb water when exposed to high relative humidities. This can lead quite rapidly to the complete dissolution of the original minerals. Hydration and hydrolysis can result in *in-situ* alteration of minerals, although fortunately these particular reactions are rare in common rock-forming minerals.

A relative humidity range of 40–60% is tolerable for general storage conditions, with a norm of about 50%. The most common reason for exceeding this is that core is stored wet, and in this case relative humidities in individual boxes may approach 100%. This will not be cured by any amount of atmospheric control, and therefore core should be thoroughly dried (in equilibrium with air at around 50% relative humidity) before boxes are closed.

If dried core is stored in closed boxes, it will be protected from short-term increases in relative humidity above about 60% (as occurs during some British summers), but core stores in a more permanently humid climate should ideally be fitted with dehumidifiers.

Low relative humidities (below 40%) occur during British winters and in drier climates, but are also caused by many types of electrical air heater. Again, short-term fluctuations are buffered by closed boxes (especially those of stout wood), but the long-term solution to the problem is the installation of humidifiers.

The other main environmental problem in core stores is the settling of airborne dust. Although some of this originates from general atmospheric pollution, especially in urban areas, the majority is created within the core store itself. Dust should be prevented from reaching the cores themselves by keeping them in closed boxes, and can be reduced by good housekeeping. This includes keeping rock preparation facilities (such as rock saws) separate from core storage areas, and keeping the core store itself as clean as possible. Few core stores will have the resources for regular dusting down of individual core boxes, but at least the floors should be kept clean so that dust is not constantly raised by the passage of feet and trolleys. When boxes are removed for study of cores, the opportunity should not be missed to dust, if necessary, both the boxes themselves, and the spaces they have temporarily vacated in the racking.

Although some core is frozen at the wellsite in order to preserve pore fluids or to prevent disaggregation, long-term storage of frozen core in large quantities is not normally viable for economic reasons. However, it is not uncommon to keep some types of core in cold storage, just above freezing point. This is a useful option for unconsolidated wet, clay-rich cores (such as seabed samples), which, because of their high organic content, tend to decompose rapidly (with attendant smells) at room temperature, before cracking and shrinking on drying. The core is normally sealed in 'lay-flat' polythene tubing or similar material, and kept in containers that do not lose strength when wet.

The above comments on core store facilities assume that a balance needs to be struck between core preservation requirements and economic considerations. The great bulk of cores stored in these conditions are likely to remain in a reasonable state for many years, and indeed few core stores meet all the criteria suggested. However, these are not museum conditions, and it is not guaranteed that there will be no deterioration whatsoever. If the long-term preservation of particular sections of core in a pristine state is crucial, environmental conditions will need to be much more tightly controlled.

7.2 Consolidated core

Dry consolidated core is not generally subject to major mechanical damage, but it must be kept clean, in correct order, and in a neat package for ease of handling and storage. Closed wooden boxes with lids, of the kind sometimes used at wellsite (Section 3.2.1), are ideal for storage, since they are sturdy, and the thick wood provides a useful buffer against short-term changes in air temperature and humidity. They are rather heavy, though, and unless a forklift truck or other mechanical lifting aid is available, wooden boxes are difficult and potentially dangerous to move when loaded with core of more than, say, 5 cm diameter. They can also be rather bulky, being formed of wood up to 2 cm thick, and thus increase storage costs. Stout cardboard boxes are sometimes used, but they tend to sag badly and even collapse when heavily loaded. They also become weak and misshapen when wet. Core boxes of heavy-duty corrugated plastic are available, and although not as solid as wooden boxes, they are light and are a good compromise where wood is considered to be too heavy or bulky.

Lids of wooden boxes may be screwed on, but this is a cumbersome process, especially if the core is to be viewed other than very rarely. Hinged lids may also be used, but this is liable to be expensive if large volumes of core are being stored. If the lids fit on closely, in such a way that they will not slip off easily, they may not need to be fastened at all within the confines of the core store. Otherwise, several turns of heavy-duty tape around the box at each end will usually be sufficient.

It is best to adopt some method of storing the core in the box that prevents either the core from rolling about, or the individual pieces of fragmented core from becoming disordered. Some core boxes fit the core so neatly that it will not move, but the same core can be extremely difficult to remove from the box when required.

Rags are sometimes stuffed down the side, and are effective, although they do tend to attract damp and dust and can cover a core in fluff. Foam rubber sheets may also be used, but they are harder to mould to the shape of the core, and they do deteriorate in time. If half-core is being stored, it may be sufficient to lay it flat side down in the box, but this can be a nuisance for the geologist who has to turn over each piece to view the slabbed face. More elegant solutions tend to add to the curating cost and time. If the box has a long gutter-like depression down the centre of its length, the core will tend to remain in place. This can be formed by actual plastic guttering material mounted firmly in the base of the box, with a diameter equal to or marginally greater than that of the core, or by expanded polystyrene or foam inserts, which can be designed to fit in the box with the same effect. This arrangement can hold the core quite firmly, with the slabbed face (if any) uppermost.

It is useful to have some packing on top of the core to prevent it from jumping around, especially if the box is accidentally dropped. Foam rubber sheets cut to the shape of the box can be used. Whichever type of box is used, it should be marked clearly on the outside with well number or name, core number, and the depth of the top and bottom of the core contained in the box.

Other labels may of course need to be added for curating purposes. In addition to ordinary core or half-core, some core may have been specially treated, such as having been mounted in clear resin for protection and ease of handling (Section 3.3.3). These are generally thin slabs, and are much easier to deal with than heavier half-core. If sampling is not required, they can simplify the job of examining core, and are not easily damaged. Being flat and light they are easier to store than half-core, and can conveniently be stacked in shallow wooden trays, only the top one of which needs a lid.

Unfortunately, these resinated slabs do sometimes warp with age, and the long-term stability of the resin (over many years to decades) has not yet been sufficiently tested. In the medium term, at least, they are a great benefit both to the core store curator and to the geologist.

7.3 Unconsolidated core

Unconsolidated core can be considered under two categories. There is core that is intact but very friable, which will break into its constituent particles with rough handling; and there is core that just consists of piles of loose sand, pebbles or similar components.

Unconsolidated cores can be preserved indefinitely by mounting in clear resin, as with consolidated cores, in which case they may be treated like other resinated slabs. These are of considerable value for core description purposes, but the range of analyses that may be undertaken on core impregnated with resin is severely limited, and it is still important to be able to preserve core as well as possible in its original state.

Unless friable core is of sufficient value to consider freezing—this is normally not the case except while it is actively being worked on—it can be packed little differently from consolidated core. Special care should be taken to ensure that it cannot rock around in the box, and any packing materials should exert only light pressure on the core. It may be best to lay the core within a gutter depression into which it fits closely, so that no additional packing is required at all. In this event it is vital to ensure that the box is kept upright and handled gently at all times. A note in large letters on the outside may help: WITH CARE – FRIABLE CORE, and all core store personnel must be aware that this means what it says!

Precautions should be taken to ensure that, if friable core does begin to disintegrate, as it always will to some degree, the resultant debris is contained within the box, as near to in-place as possible. If the core is liable to collapse into a fine sand, there should be no hole in the base of the box through which it might drain. There are stories of such core having been transported in slatted wooden boxes that, on arrival at their destination, have been found to be almost empty, the core having been left behind as a trail of loose sand along the road.

Wholly disaggregated core is generally simpler to store than friable but intact core, since it can suffer no further damage, although it may become mixed up. It is still important to keep the core in its proper order as much as possible. It may be laid out in a gutter depression within a core box, but this is unsuitable if the box is liable to be tilted or upturned.

An alternative is to pack the core into individually marked sample bags, laid out in order within the core boxes. Each bag might contain, say, 15 cm of core. This is generally satisfactory, but it makes the core difficult to inspect properly. There is really no wholly satisfactory method of storing disaggregated core. Any method used will be a compromise, and that chosen will depend on the priorities of any particular situation.

7.4 Core samples

Before the core reaches the core store for long-term curation, it may have been sampled on one or more occasions. Some samples may have been wholly consumed by the analytical process, others will have been altered in some way (for example, made into thin sections) and will be curated separately, whereas yet other core chips will be virtually unaltered and may be reunited with the core. If the procedure suggested in Section 3.2.5 has been followed, a small card will have been left in the core box at each point where a sample has been taken. These cards should of course have been transferred to any new core box on curation, and will help to ensure that samples are returned to their correct position.

It will sometimes be useful to trace at a later date the actual core chip used in a particular analysis. Furthermore, it is possible that sample chips may have suffered some slight contamination. For these reasons, the sample may be returned to the core box, but may best be left in a separate sample bag appropriately marked.

The same procedure applies to larger samples, such as lengths of core preserved in wax. In this case, however, so long as the samples remain waxed there will be no advantage in placing them in separate bags, since their nature will be apparent.

Appendix 1 Standard symbols and abbreviations

about	abt	bentonite(-ic)	bent
above	ab	bitumen(-inous)	bit
absent	abs	bioclastic	biocl
abundant	abd	bioherm(-al)	bioh
acicular	acic	biomicrite	biomi
agglomerate	aglm	biosparite	biosp
aggregate	agg	biostrome(-al)	biost
algae, algal	alg	biotite	biot
allochem	allo	birdseye	bdeye
altered	alt	bivalve	biv
alternating	altg	black(-ish)	blk, blksh
ammonite	amm	blade(-ed)	bld
amorphous	amor	blocky	blky
amount	amt	blue(-ish)	bl, blsh
and	&	bored(-ing)	bor
angular	ang	bottom	btm
anhedral	ahd	botryoid(-al)	bot
anhydrite(-ic)	anhy	boulder	bld
anthracite	anthr	boundstone	bdst
aphanitic	aph	brachiopod	brach
appears, apparently	ap	brackish	brak
approximate	apprx	branching	brhg
aragonite	arag	break	brk
arenaceous	aren	breccia(-ted)	brec
argillaceous	arg	bright	brt
arkose(-ic)	ark	brittle	brit
as above	a.a.	brown	brn
asphalt(-ic)	asph	bryozoa(-n)	bry
assemblage	assem	bubble	bubl
associated	assoc	buff	bu
at	@	burrow(-ed)	bur
authigenic	authg		
average	av	calcarenite	clcar
		calcilutite	clclt
band(-ed)	bnd	calcirudite	clcrd
basalt(-ic)	bas	calcisiltite	clslt
basement	bm	calcisphere	clcsp
become(-ing)	bcm	calcite(-ic)	calc, calctc
bed(-ded)	bd	calcareous	calc
bedding	bdg	caliche	cche
belemnite	belem	calcrete	ccte

carbonaceous	carb
carbonized	cb
cast	cst
cavern(-ous)	cav
caving	cvg
cement(-ed, -ing)	cmt
cephalopod	ceph
chalcedony(-ic)	chal
chalk(-y)	chk, chky
chert(-y)	cht
chitin(-ous)	chit
chlorite(-ic)	chlor
chocolate	choc
circulate(-ion)	circ
clastic	clas
clay(-ey)	cl
claystone	clst
clean	cln
clear	clr
cleavage	clvg
closed	clsd
cluster	clus
coal	coal
coarse	crs
coarsening-up	c.u.
coated(-ing)	cotd, cotg
coated grains	cotd gn
cobble	cbl
colour(-ed)	col
common	com
comminuted	comm
compact	cpct
compare	cf
concentric	cncn
conchoidal	conch
concretion(-ary)	conc
conglomerate(-ic)	cgl
considerable	cons
consolidated	consol
conspicuous	conspic
contact	ctc
contamination(-ed)	contam
content	cont
contorted	cntrt
coquina(-oid)	coq, coqid
coral, coralline	cor, corln
core	core, c
covered	cov
cream	crm
crenulated	cren
crinkled	crnk
crinoid(-al)	crin, crinal
cross	x
cross-bedded	x-bd
cross-laminated	x-lam
cross-stratified	x-strat
crumpled	crpld
cryptocrystalline	crpxln
crystal(-line)	xl, xln
cube, cubic	cub
cuttings	ctgs
dark(-er)	dk, dkr
dead	dd
debris	deb
decrease(-ing)	decr
dense	dns
description	descr
desiccation	desicc
detrital	detr
devitrified	devit
diagenesis(-etic)	diagn
diameter	dia
disseminated	dissem
dish	dsh
ditto	" or do
dolerite	doler
dolomite(-ic)	dol
dominant(-ly)	dom
drilling	drlg
drusy	dru
earthy	ea
east	E
echinoid	ech
elevation	elev
elongate	elong
embedded	embd
equant	eqnt
equivalent	equiv
euhedral	euhd
euxinic	eux
evaporite(-itic)	evap
excellent	ex
exposed	exp
extraclast(-ic)	exclas
extremely	extr

extrusive rock, extrusive	exv
facet(-ed)	fac
facies	fcs
faint	fnt
fair	fr
fault(-ed)	flt
fauna	fau
feet	Ft
feldspar(-athic)	fspr
fenestra(-al)	fen
ferruginous	ferr
fibrous	fibr
fine(-Iy)	f,fnly
fining-up	f.u.
fissile	fis
flaggy	flg
flake, flaky	flk
flaser	flas
flat	fl
floating	fltg
flora	flo
fluorescence(-ent)	fluor
foliated	fol
foot	Ft
formation	Fm
fossil(-iferous)	foss
fracture(-d)	frac
fragment(-al)	frag
frequent	freq
fresh	frs
friable	fri
fringe(-ing)	frg
frosted	fros
frosted quartz grains	F.Q.G
gabbro	gab
gastropod	gast
gas	gas
generally	gen
geopetal	gept
glass(-y)	glas
glauconite(-itic)	glauc
gloss(-y)	glos
gneiss(-ic)	gns
good	gd
grading	grad
grain(-s, -ed)	gr
grainstone	grst
granite	grt
granite wash	G.W.
granule(-ar)	gran
grapestone	grapst
graptolite	grap
gravel	grv
grey(-ish)	gry, grysh
greasy	gsy
green(-ish)	gn, gnsh
gypsum(-iferous)	gyp
hackly	hkl
halite(-iferous)	hal
hard	hd
haematite(-ic)	haem
heavy	hvy
heterogeneous	hetr
high(-1y)	hi
homogeneous	hom
horizontal	hor
hydrocarbon	hydc
igneous rock (igneous)	ig
imbricated	imbr
impression	imp
inch	In
inclusion(-ded)	incl
increasing	incr
indistinct	indst
indurated	ind
in part	I.P.
insoluble	insl
interbedded	intbd
intercalated	intercal
intercrystalline	intxln
intergranular	intgran
intergrown	intgn
interlaminated	intrlam
interparticle	intpar
interval	intvl
intraclast(-ic)	intclas
intraparticle	intrapar
intrusive rock, intrusive	intr
invertebrate	invtb
iridescent	irid
ironstone	Fe-st
irregular(-ly)	irr
isopachous	iso

jasper	jasp
joint(-ed, -ing)	jt
kaolin(-itic)	kao
lacustrine	lac
lamina(-tions, -ated)	lam
large	lge
laterite(-itic)	lat
layer	lyr
leached	lchd
lens, lenticular	len, lent
light	lt
lignite(-itic)	lig
limestone	lst
limonite	lim
limy	lmy
lithic	lit
lithology(-ic)	lith
little	ltl
littoral	litt
local	loc
long	lg
loose	lse
lower	lwr
luster	lstr
lutite	lut
macrofossil	macrofos
magnetite, magnetic	mag
manganese, manganiferous	Mn, mn
marble	mbl
marl(-y)	mrl
marlstone	mrlst
marine	marn
maroon	mar
massive	mass
material	mat
matrix	mtrx
maximum	max
medium	med
member	Mbr
meniscus	men
metamorphic rock	meta
metamorphic(-osed)	meta, metaph
metre	m
mica(-ceous)	mic
micrite(-ic)	micr
microcrystalline	microxln
microfossil(-iferous)	microfos
micrograined	micgr
micropore(-osity)	micropor
microspar	micropr
microstylolite	microstyl
middle	mid
milky	mky
mineral(-ized)	min
minor	mnr
moderate	mod
mould(-ic)	mol
mollusc	moll
mosaic	mos
mottled	mott
mud(-dy)	md, mdy
mudstone	mdst
muscovite	musc
nacreous	nac
nodules(-ar)	nod
north	N
no sample	n.s.
no visible porosity	n.v.p
numerous	num
occasional	occ
ochre	och
oil	oil
olive	olv
ooid(-al)	oo
oolite(-itic)	ool
oncolite(-oidal)	onc
opaque	op
open	opn
orange(-ish)	or, orsh
organic	org
orthoclase	orth
orthoquartzite	o-qtz
ostracod	ostr
overgrowth	ovgth
oxidized	ox
packstone	pkst
part(-ly)	pt
particle	par
parting	ptg
parts per million	ppm
patch(-y)	pch

pebble(-y)	pbl
pellet(-al)	pel
pelletoid(-al)	peld
permeability(-able)	perm, k
petroleum, petroliferous	pet
phlogopite	phlog
phosphate(-atic)	phos
phyllite, phyllitic	phyl
phreatic	phr
pink	pk
pinkish	pkish
pin-point(porosity)	p.p.
pisoid(-al)	piso
pisolite, pisolitic	pisol
pitted	pit
plagioclase	plag
plant	plt
plastic	plas
platy	plty
polish, polished	pol
polygonal	poly
porcelanous	porcel
porosity, porous	por, ø
possible(-ly)	poss
predominant(-ly)	pred
preserved	pres
preserved sample	p.s.
primary	prim
probable(-ly)	prob
production	prod
prominent	prom
pseudo–	ps
pumice-stone	pst
purple	purp
pyrite(-itized; -itic)	pyr
pyroclastic	pyrcl
quartz(-ose)	qtz
quartzite(-ic)	qtzt
radial(-ating)	rad
radiaxial	radax
range	rng
rare	r
recemented	recem
recovery(-ered)	rec
recrystallized	rexlzd
red(-dish)	rd, rdsh
reef(-oid)	rf
remains	rems
replaced(-ment)	rep
residue(-ual)	res
resinous	rsns
rhomb(-ic)	rhb
ripple	rpl
rock	rk
rootlet	rtlt
round(-ed)	rnd, rndd
rounded, frosted, pitted	r.f.p.
rubble(-bly)	rbl
saccharoidal	sacc
salt(-y)	sa
salt and pepper	s & p
salt water	S.W.
same as above	a.a.
sample	spl
sand(-y)	sd, sdy
sandstone	sst
saturation(-ated)	sat
scarce	scs
scattered	scat
schist(-ose)	sch
scour(-ed)	scr
secondary	sec
sediment(-ary)	sed
selenite	sel
SEM sample	SEM
shale(-ly)	sh
shell(-y)	shl, shly
show	shw
siderite(-itic)	sid
sidewall core	S.W.C.
silica(-iceous)	sil
silky	slky
silt(-y)	slt, slty
siltstone	sltst
similar	sim
size	sz
skeletal	skel
slabby	slb
slate(-y)	sl
slickenside(-d)	slick
slight(-1y)	sli, slily
slump(-ed)	slmp
small	sml
smooth	sm

soft	sft
solution, soluble	sol
somewhat	smwt
sorted(-ing)	srt, srtg
south	S
spar(-ry)	spr
sparse(-Iy)	sps, spsly
speck(-led)	spk, spkld
sphalerite	sphal
spherule(-itic)	spher
spicule(-ar)	spic
splintery	splin
spotted(-y)	sptd, spty
stain(-ed, -ing)	stn
stalactitic	stal
strata(-ified)	strat
streak(-ed)	strk
striae(-ted)	stri
stringer	strgr
stromatolite(-itic)	stromlt
structure	str
stylolite(-itic)	styl
subangular	sbang
sublithic	sblit
subrounded	sbrndd
sucrosic	suc
surface	surf
syntaxial	syn
tabular(-ate)	tab
tan	tn
terrigenous	ter
texture(-ed)	tex
thick	thk
thin	thn
thin-bedded	t.b.
thin section sample	T.S.
throughout	thr
tight	ti
top	tp
total depth	T.D.
tough	tgh
trace	tr
translucent	trnsl
transparent	trnsp
trilobite	tril
trough	trgh
tube(-ular)	tub
tuff(-aceous)	tf
type(-ical)	typ
unconformity	unconf
unconsolidated	uncons
underlying	undly
uniform	uni
upper	u
vadose	vad
variation(-able)	var
variegated	vgt
varicoloured	varic
varved	vrvd
vein(-ing, ed)	vn
veinlet	vnlet
vertebrate	vrtb
vertical	very
very	v
very poor sample	V.P.S.
vesicular	ves
violet	vi
visible	vis
vitreous(-ified)	vit
volatile	volat
volcanic rock	volc
volcanic	vug
wackestone	wkst
water	wtr
wavy	wvy
waxy	wxy
weak	wk
weathered	wthd
well	wl
west	W
white	wh
with	w
without	w/o
wood	wd
XRD sample	XRD
yellow(-ish)	yel, yelsh
zircon	zr
zone	zn

LITHOLOGIES

Clay/Claystone/Mudstone

Shale

Silt/Siltstone

Sand/Sandstone

Gravel/Conglomerate (Extraformational)

Conglomerate (Intraformational)

Breccia

Limestone

Dolomite

Anhydrite/Gypsum

Halite

Coal

Tuff/Ash

Extrusive Igneous

Intrusive Igneous

Metamorphic (undifferentiated)

Gneiss

Schist

ACCESSORIES

▲ Chert/Chalcedony

Gl Glauconite

Mc Mica

Ph Phosphate

Py Pyrite

Sid Siderite

N.B. Abbreviations from the standard list may additionally be used

QUALIFIERS

----- Argillaceous

Bit Bituminous

Calcareous

Carbonaceous

(Py) Concretionary (with qualifier indicating mineralogy)

Dolomitic

Microfossiliferous

Mud-clast bearing

Oolitic

Fossiliferous (i.e. stylized representation of type)

Woody debris

◊ Vuggy

M W P G B — Lime mud, wackestone, packstone, grainstone, boundstone, used with carbonate rocks according to Dunham classification

SEDIMENTARY STRUCTURES

Massive/No apparent bedding (or left blank)
Indistinct/Uncertain bedding (or other structure as appropriate)
Horizontal bedding
Cross-bedding
Lamination
Imbrication
Cross-lamination
Flaser bedding
Lenticular bedding
Wavy bedding
Dish structures
Wave ripples
Bioturbation
Burrowing (stylized representation of type)
Rootlet traces
Stylolite
Mud flakes/curls
Desiccation cracks
Flute casts
Load casts
Geopetal structure
Injection structure
Flame structure
Fluidization structure
Slumping

BED BOUNDARIES

Sharp
Scoured/Erosional
Gradational (or shown by merging of lithology ornaments)

OTHER STRUCTURES

Qtz — Closed – showing infill
2mm — Open fractures – showing separation
Slk — Slickensides
Microfaults
Downward increase in grain size

SAMPLE AND DATA POINTS

Pl — Photograph
TS, Pal, G, Ø — Sample point for thin section, palynology, geochemistry, porosity etc.
Recovered / Not recovered — Sidewall core point
SO 060° N — Orientation of scribe line
P.S. — Preserved sample
Core missing/not recovered
Extent of core recovery, with core number

Appendix 2 Equipment for wellsite coring operations and core logging

The equipment needed will obviously depend on the precise nature of the operation. The following checklists are intended as a basis, against which additions and deletions may be made.

Wellsite operations

- Core catching boxes or laying-out trays/tables (these need to be of a greater length than expected coring length, to allow for an initially loose arrangement of core)
- Plastic lay-flat tubing with heat-sealing machine or staple gun (with staples)
- Plastic or linen sample bags for broken core and samples
- Cling film, aluminium foil etc.
- Wax and wax bath, or alternative for sealing core
- Nylon reinforced adhesive tape, for sealing core boxes, and/or screws/ nails as appropriate
- Oil and waterproof marker pens and/or wax crayons
- Measuring tape
- Geological hammer
- Cutting knife
- Report sheets on clipboard, or hardback notepad
- Core packing materials—boxes and rags, foam etc.

Core logging

- Core logging sheets on clipboard, or hardback notepad
- Pencil with sharpener, or fine-tipped water–proof pen
- Ruler
- Eraser
- Hand lens and/or binocular microscope

- Dilute HCl in dropper bottle
- Steel point (to gauge mineral hardness)
- Grain-size chart
- Polythene spray bottle with water
- Supply of water and sponge or cloth for washing core
- Blunt knife for scraping mud from core

Appendix 3 Standard core barrel sizes

Outside the hydrocarbon industry, most core barrels are designed in a standard range of external diameters, denoted by a letter. The external diameter is equivalent to the hole diameter, assuming that no erosion or caving of the borehole wall occurs. The diameter of the core cut is of course less than this. The standard sizes are as follows:

Size symbol	Nominal hole diameter (mm)	Core diameter: Standard types (mm)	Core diameter: Thin-walled (T series)
R	30	—	19
E	38	21	23
A	48	30	33
B	60	42	44
N	76	55	59
H	99	76	81
P	120	92	—
S	146	113	—
U	175	140	—
Z	200	165	—

In addition to a letter denoting the core barrel diameter, a second letter is often used to indicate the core barrel design. Common designs are **F** (face discharge core barrel), **G** (general design), **M** (designed for friable lithologies) and **T** (a thin-walled barrel which cuts a slightly larger-diameter core).

Core barrel diameters in the hydrocarbon industry are stated in inches. The metric equivalents are given in the following table for information:

External diameter		Internal diameter	
Inches	mm	Inches	mm
4¾	120.7	2 ⅝	66.7
5¾	146.1	3 ½	88.9
6¾	171.5	4	101.6
7¼	184.2	4	101.6
8	203.2	5¼	133.4
12¼	311.2	5¼	133.4

Although all of the above sizes are common, core barrels are also manufactured in numerous other sizes.

Bibliography

The following are a few references that enlarge on various aspects of core handling and analysis.

Anderton, R. (1985) Clastic facies models and facies analysis. In *Sedimentology: Recent Developments and Applied Aspects*, Brenchley, P.J. and Williams, B.P.J. (eds). Geological Society Special Publication No. 18.

This is a short but incisive critique on the nature and role of facies models for interpreting geological sequences.

British Standards Institution (1999) *Code of practice for site investigations* (BS 5930:1999), 193 pp.

A useful guide (and for some an obligatory code of practice) for engineers and geotechnical specialists.

Brunton, C.H.C., Besterman, T.P. and Cooper, J.A. (1985) *Guidelines for the curation of geological materials*. Geological Society Miscellaneous Paper No. 17.

This reference includes useful information on the long-term storage of rock samples, in addition to sample handling, preparation and photography.

Carter, M. (1983) *Geotechnical Engineering Handbook*. Pentech Press Ltd, Plymouth, 226 pp.

A useful guide including methods of coring and core testing for geotechnical purposes.

Deere, D.U., Hendron, A.J., Patton, F.D. and Cording, F.J. (1967) Design of surface and near surface construction in rock. *Proceedings of the 8th Symposium on Rock Mechanics, University of Minnesota*, 237–302.

Defines the rock quality designation (RQD).

Ege, J.R. (1968) Stability index for underground structures in granitic rock, in Nevada Test Site. *Memoirs of the Geological Society of America* **110**, 185–198.

Defines the rock stability index.

Exlog (1985a) *Coring Operations*. Reidel, Dordrecht, 174 pp.

An excellent industry training manual of core sampling and analysis techniques in the petroleum industry.

Exlog (1985b) *Field Geologists' Training Guide*. Reidel, Dordrecht, 291 pp.

'Field geology' refers here to wellsite geology in the petroleum industry. This is a dated but still useful account of deep drilling techniques, including all aspects of coring operations.

Goddard, E.N, Trask, P.D, De Ford, R.K., Rove, O.N., Singewald, J.T. and Overbeck, R.M. (1948) *Rock-Color Chart*. Geological Society of America, Colorado.

The standard colour chart for geologists.

Knill, J.L., Cratchley, C.R., Early, K.R, Gallois, R.W., Humphreys, J.D., Newbery, J., Price, D.G. and Thurrell, R.G. (1970) The logging of rock cores for engineering purposes. *Quarterly Journal of Engineering Geology* **3**, 1–24.

Recommendations from a working party of the Geological Society Engineering Group, widely adopted by engineering geologists.

Posamentier, H. and Walker, R. (2006) *Facies Models Revisited.* SEPM Special Publication No. 84 (CD). Society for Sedimentary Geology, Tulsa, OK.

An authoritative summary of sedimentary facies models.

Reading, H.G. (1996) *Sedimentary Environments: Processes, Facies and Stratigraphy*, 3rd edition. Blackwell Science, Oxford, 688 pp.

A thorough and authoritative account of modern sedimentological and facies models.

Rider, M.H. (1986) *The Geological Interpretation of Well Logs.* Blackie, Glasgow, 175 pp.

A useful and user-friendly introduction to electric wireline logs.

Rothwell, R.G. (ed.) (2006). *New Techniques in Sediment Core Analysis.* Geological Society Special Publication No. 267, 272 pp.

A good summary of some modern and unconventional analytical techniques for use on core.

Schlumberger (1991) *Log Interpretation Principles/Applications.* Schlumberger Educational Services, Houston, 198 pp.

The authoritative introduction to electric wireline logs by the world's largest wireline logging company.

Swanson, R.G. (1981) *Sample Examination Manual.* Methods in Exploration Series, American Association of Petroleum Geologists.

Developed from the Shell Oil Co. standard sample examination manual, and containing details of analytical techniques (biased towards oil industry requirements) and various useful charts, legends and abbreviations.

Worthington, A.E., Gidman, J. and Newman, G.H. (1987) Reservoir petrophysics of poorly consolidated rocks, I: Wellsite procedures and laboratory methods. *Proceedings of the SPWLA 28th Annual Logging Symposium, London*, Paper BB.

A useful guide to handling unconsolidated cores at wellsite, with the first description of wellsite resination techniques similar to those currently practised, as developed by the Chevron Petroleum Company.

Index

Numbers in **boldface** refer to text figures.